Daniel Lanzerstorfer

Global Instabilities of Incompressible Plane-Channel Flows

Daniel Lanzerstorfer

Global Instabilities of Incompressible Plane-Channel Flows

The Backward-Facing-Step, Forward-Facing-Step and Sudden-Expansion Problems

Südwestdeutscher Verlag für Hochschulschriften

Impressum/Imprint (nur für Deutschland/only for Germany)
Bibliografische Information der Deutschen Nationalbibliothek: Die Deutsche Nationalbibliothek verzeichnet diese Publikation in der Deutschen Nationalbibliografie; detaillierte bibliografische Daten sind im Internet über http://dnb.d-nb.de abrufbar.
Alle in diesem Buch genannten Marken und Produktnamen unterliegen warenzeichen-, marken- oder patentrechtlichem Schutz bzw. sind Warenzeichen oder eingetragene Warenzeichen der jeweiligen Inhaber. Die Wiedergabe von Marken, Produktnamen, Gebrauchsnamen, Handelsnamen, Warenbezeichnungen u.s.w. in diesem Werk berechtigt auch ohne besondere Kennzeichnung nicht zu der Annahme, dass solche Namen im Sinne der Warenzeichen- und Markenschutzgesetzgebung als frei zu betrachten wären und daher von jedermann benutzt werden dürften.

Coverbild: www.ingimage.com

Verlag: Südwestdeutscher Verlag für Hochschulschriften GmbH & Co. KG
Heinrich-Böcking-Str. 6-8, 66121 Saarbrücken, Deutschland
Telefon +49 681 37 20 271-1, Telefax +49 681 37 20 271-0
Email: info@svh-verlag.de

Approved by: Wien, TU, Diss., 2012

Herstellung in Deutschland (siehe letzte Seite)
ISBN: 978-3-8381-3370-6

Imprint (only for USA, GB)
Bibliographic information published by the Deutsche Nationalbibliothek: The Deutsche Nationalbibliothek lists this publication in the Deutsche Nationalbibliografie; detailed bibliographic data are available in the Internet at http://dnb.d-nb.de.
Any brand names and product names mentioned in this book are subject to trademark, brand or patent protection and are trademarks or registered trademarks of their respective holders. The use of brand names, product names, common names, trade names, product descriptions etc. even without a particular marking in this works is in no way to be construed to mean that such names may be regarded as unrestricted in respect of trademark and brand protection legislation and could thus be used by anyone.

Cover image: www.ingimage.com

Publisher: Südwestdeutscher Verlag für Hochschulschriften GmbH & Co. KG
Heinrich-Böcking-Str. 6-8, 66121 Saarbrücken, Germany
Phone +49 681 37 20 271-1, Fax +49 681 37 20 271-0
Email: info@svh-verlag.de

Printed in the U.S.A.
Printed in the U.K. by (see last page)
ISBN: 978-3-8381-3370-6

Copyright © 2012 by the author and Südwestdeutscher Verlag für Hochschulschriften GmbH & Co. KG and licensors
All rights reserved. Saarbrücken 2012

"Now I think hydrodynamics is to be the root of all physical science, and is at present second to none in the beauty of its mathematics."

William Thomas (Lord Kelvin) to George G. Stokes, 1857.

Acknowledgements

The permission of the Cambridge University Press to re-use the material, originally written for publications in the Journal of Fluid Mechanics, is gratefully acknowledged!

Reprinted with permission.

Contents

1. **Motivation and Introduction**   1

2. **Mathematical Formulation**   6
   - 2.1. Basic-Flow Equations 6
   - 2.2. Linear Stability Analysis 7
   - 2.3. Energy Analysis 10
   - 2.4. Adjoint Analysis 12

3. **Numerical Implementation**   14
   - 3.1. Newton's Method 14
   - 3.2. Finite-Volume Discretization 18
   - 3.3. Grid Generation 21
   - 3.4. Eigenvalue-Detection Strategies 23
   - 3.5. Algorithms for Root-Finding and Minimization 25
   - 3.6. Plane Poiseuille Flow 29

4. **Results**   31
   - 4.1. The Backward-Facing-Step Problem 31
     - 4.1.1. Problem Formulation 31
     - 4.1.2. Scientific Background 32
     - 4.1.3. Results 34
     - 4.1.4. Conclusion 55
   - 4.2. The Forward-Facing-Step Problem 57
     - 4.2.1. Problem Definition 57
     - 4.2.2. Scientific Background 57
     - 4.2.3. Results 59
     - 4.2.4. Conclusion 69
   - 4.3. The Plane Sudden-Expansion Problem 71
     - 4.3.1. Problem Formulation 71
     - 4.3.2. Scientific Background 72
     - 4.3.3. Results 74

4.3.4. Conclusion	93

5. Summary and Outlook   **95**

A. Derivation of the Reynolds–Orr Equation   **97**

B. Jacobian-Free Newton–Krylov Approach   **99**

Bibliography   **102**

1. Motivation and Introduction

The separation of the flow as it passes over a sharp corner is of practical as well as theoretical interest. Separated/reattached flows appear in many industrial applications such as diffusers, aerofoils or turbine blades. The phenomenon of flow separation often leads to drastic losses in aerodynamic performances of aerofoils or automotive vehicles. Many aspects of such flows can be studied in ideal geometries such as the backward-facing step, forward-facing step and plane sudden expansion. These geometries are simple as there are only a few geometric parameters, but not necessarily simple in terms of flow phenomena. The systems in question consist of plane channels, which exhibit sudden expansions/constrictions in the form of steps. Figure 1.1 shows as an example the visualization of the experimental flow of the sudden-expansion problem. The geometries consist of a plane inlet channel, followed by a suddenly expanded (constricted) channel. The direction of the bulk flow is the x direction and the steps are vertically aligned in the y direction, perpendicular to the x axes.

Regardless of the step heights, the flow is characterized by regions of separated flow in the form of more or less strained vortices immediately behind the steps. They are referred to as primary vortices because they are the first ones to appear as the Reynolds number Re is increased. For higher Reynolds numbers, secondary and higher-order regions of separation arise further downstream, alternatingly located on both walls of the channel. The experimental investigations of the two-dimensional flow reveal that within the laminar regime the length of the primary vortices increases linearly with Re. For very large step heights, a thin plane jet emerges from the opening of the inlet channel and the streamlines of the primary vortices are almost circular near its centre. When the step heights are reduced, these vortices are getting

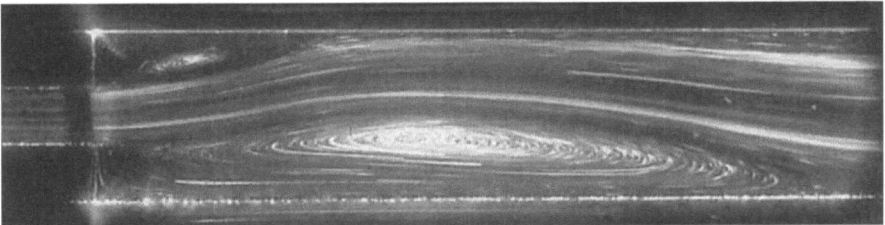

Figure 1.1.: Experimental flow visualization of the sudden-expansion problem (Fearn et al., 1990).

more and more strained and elongated in the streamwise direction. For very small step heights, the bulk flow is almost parallel and a relatively high Reynolds number is required to destabilize the flow. The experimental findings show that the characteristics of the vortices immediately behind the steps are of key importance for the flow instabilities as the recirculation bubbles provide a feedback for the perturbations. Near the outflow region, a plane Poiseuille flow is realized again if the Reynolds number is sufficiently small and the outflow channel is long enough. Also the inlet channel has to be sufficiently long such that a plane Poiseuille flow is realized in the experiments.

In fluid mechanics, one is, amongst other things, interested in the question which flow structure will be realized for certain initial and boundary conditions. Any persistent flow pattern, being observed in the experiments, corresponds to a stable solution of the Navier–Stokes equations. Here stability refers to a reference state, which is called the basic flow. In order to understand the above-mentioned flow patterns and phenomenologies, in the present work the basic flows are studied under idealized conditions, where a fully developed laminar plane Poiseuille flow is prescribed within the inlet channel. Moreover, the basic flows are time-independent, two-dimensional and are assumed to be infinitely extended in the homogeneous spanwise (z) direction. This translation symmetry allows one to study intrinsic effects, unmasked by sidewall perturbations. Thus it is interesting to know when the basic flow loses its symmetry and gets three-dimensional by increasing the Reynolds number Re. For the purpose of hydrodynamic stability, certain perturbations are added to the steady basic flow at the time $t = 0$. The temporal evolution of the perturbations determines, whether the basic state is stable or unstable.

For small Reynolds numbers, all perturbations, independent of their amplitudes, decay in an exponential and monotonic way for $t \to \infty$. The basic flow is asymptotically monotonically stable for Re < Re_e, where Re_e defines the energy stability limit according to Joseph (1976). For Re_e < Re < Re_a, the perturbation amplitudes might grow initially $t > 0$, until they tend to zero for $t \to \infty$. In this scenario, the basic flow is only asymptotically stable as the disturbances exhibit transient growth (see figure 1.2). The amplification of the perturbations due to transient growth can be quite strong as it is the case in the plane Poiseuille flow (Trefethen et al., 1992, 1993), where the conservative energy stability limit accounts for $\text{Re}_e = 49.6$ (Carmi, 1969). In the plane Poiseuille flow, the transition to turbulence can be observed experimentally at Re ≈ 1000, even though the critical Reynolds number is $\text{Re}_c = 5772$. For Re < Re_c, the basic flow is conditionally or linearly stable, which means that the initial perturbations will decay for $t \to \infty$ only if they are infinitely small. Thus, the critical Reynolds number defines the bifurcation point, at which the basic flow loses for sure its stability, even for infinitesimal initial disturbances. For Re_a < Re_c, the bifurcation is subcritical like in the plane channel flow, whereas for $\text{Re}_a = \text{Re}_c$, the bifurcation is supercritical. In that case, the basic state is the only stable solution for Re < Re_c.

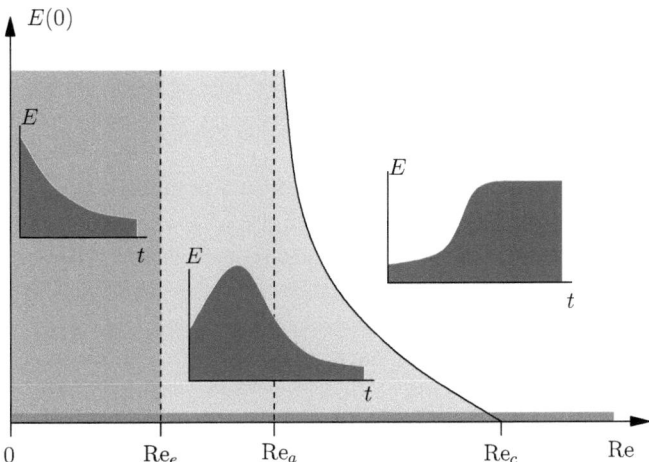

Figure 1.2.: The kinetic energy of an initial perturbation $E(t = 0)$ as measure for the amplitude of the perturbation. A basic flow is asymptotically monotonically stable for $\mathrm{Re} < \mathrm{Re}_e$, asymptotically stable for $\mathrm{Re} < \mathrm{Re}_a$ and conditionally or linearly stable for $\mathrm{Re} < \mathrm{Re}_c$. The green zone denotes the range of infinitesimal initial disturbances.

In the prevailing study, a global linear stability analysis is conducted, where the temporal behaviour of small perturbations is considered for $t \to \infty$. Large-amplitude disturbances, which could destabilize the basic flow for $\mathrm{Re} < \mathrm{Re}_c$, are not considered. Here, *global* refers to the fact that the perturbations are three-dimensional. It is worth mentioning that other approaches are possible, focusing on the spatial growth of disturbances. This ansatz is suitable for open systems, where no resonance and feedback due to closed streamlines is expected (Huerre & Rossi, 1998).

The temporal linear stability analysis is able to capture absolute as well as convective instabilities (depending on the boundary conditions). If an infinitesimal perturbation, which is initially localized in space, grows downstream as well as upstream at that fixed spatial location, then the instability is absolute. If on the other hand, the disturbance propagates as it grows in magnitude such that the perturbation ultimately decays at that fixed point in space, then the basic flow is convectively unstable, see Huerre & Monkewitz (1985). Deissler (1987) showed by considering the temporal as well as the spatial evolution of the disturbances that the instability in plane Poiseuille flow is convective in nature with the same stability boundary as obtained by Thomas (1953) and Orszag (1971), respectively. Nevertheless, in section 3.6 it is demonstrated that the critical Reynolds number $\mathrm{Re}_c = 5772$ can be determined accurately by a temporal linear stability analysis by using periodic boundary conditions in the streamwise direction.

One objective of the present work is to provide accurate reference data on the primary instabilities and its associated linear stability boundaries of the two-dimensional basic flow. These stability boundaries of great interest, as they have a severe impact on the transport and mixing properties of flows if a symmetry-breaking three-dimensional flow sets in. To date, almost all linear stability analyses for the backward-facing-step, forward-facing-step and sudden-expansion problems dealt with two-dimensional perturbations. If three-dimensional disturbances were considered (Barkley et al., 2002; Marino & Luchini, 2009), the analyses were conducted only for a few geometric parameters (step-to-main-channel heights) and for very short inlet channel lengths. In the current doctoral thesis, the three-dimensional stability analysis is carried out for a systematic variation of the geometric parameters to cover a wide range of the parameter space. Besides of that, great care is taken in order to obtain entrance-channel-length independent results.

The prevailing approach of a global stability analysis only requires to solve numerically a two-dimensional problem, as the homogeneous spanwise direction can be treated analytically. Such extensive geometric parameter variations would not be possible via a full three-dimensional simulation. Apart from the substantially increased cost of a full three-dimensional simulation, one has to bear in mind that a determination of the critical Reynolds number by simulation is prohibitively expensive, because the time-dependent simulation must be carried out for a very long time, which even diverges at the critical Reynolds number. This phenomenon, known as critical slowing down, brings about that the time scale required to obtain a saturated state tends to infinity at the critical point. Therefore, the linear stability analysis is the method of choice for accurately calculating the critical Reynolds number (if it exists).

Another objective of the present study is to shed light on the underlying physical instability mechanisms on the basis of a kinetic-energy analysis. Thereby, a relationship should be established to known instability mechanisms from simplified models. A kinetic-energy analysis provides also valuable information regarding the flow regions, where the instabilities are localized in space. These flow regions could be modified if the onset of the three-dimensional flow is to be delayed (flow control aspects (see e.g. Theofilis & Colonius, 2011)). Moreover, the numerical results are compared with previous experimental findings in order to demonstrate the physical relevance of the global instability modes found.

This work is organized in such a way that the mathematical formulations and governing equations are presented in section 2. The numerical implementations and solution strategies are described in section 3. The main part of the thesis, section 4, is dedicated to the numerical results obtained for the backward-facing step, forward-facing step and plane sudden-expansion problems. Each topic is introduced with a literature review, the problem is defined and the results are discussed, followed by a problem-specific conclusion. The appendix rounds off the present study, where details of the energy analysis and the Jacobian-free Newton–Krylov approach are described.

The results of this doctoral thesis have already been published in the Journal of Fluid Mechanics.

- LANZERSTORFER, D. & KUHLMANN, H. C. 2012a Global stability of the two-dimensional flow over a backward-facing step. *J. Fluid Mech.* **693**, 1–27.

- LANZERSTORFER, D. & KUHLMANN, H. C. 2012b Three-dimensional instability of the flow over a forward-facing step. *J. Fluid Mech.* **695**, 390–404.

- LANZERSTORFER, D. & KUHLMANN, H. C. 2012c Global stability of multiple solutions in plane sudden-expansion flow. *J. Fluid Mech.*, DOI: 10.1017/JFM.2012.184.

2. Mathematical Formulation

2.1. Basic-Flow Equations

The basic flow represents a solution of the Navier–Stokes equations, whose stability will be analysed later on. In the case of the plane channel flow, the Poiseuille flow describes the basic state. All systems treated in the present work feature a translation symmetry with respect to the homogeneous spanwise (z) direction. Therefore, the two-dimensional basic flow is considered in the (x, y) plane.

The flow of an incompressible and Newtonian fluid is governed by the Navier–Stokes and continuity equations

$$\partial_t^* \boldsymbol{u}_0^* + \boldsymbol{u}_0^* \cdot \nabla^* \boldsymbol{u}_0^* = -\frac{1}{\rho} \nabla^* p_0^* + \nu \Delta^* \boldsymbol{u}_0^*, \tag{2.1.1a}$$

$$\nabla^* \cdot \boldsymbol{u}_0^* = 0, \tag{2.1.1b}$$

where the density ρ and viscosity ν are considered constant. Viscous heating does not play an important role in the present dynamics and thus a coupling between the energy and momentum equations is not incorporated. Note that the assumption of incompressibility is valid for many fundamental fluids, such as water, air, gas, or liquid metals. A fluid is regarded as incompressible, if its density does not change with pressure variations (see e.g. Spurk, 1997). In (2.1.1a) $\nu = \mu/\rho$ represents the kinematic viscosity, where μ stands for the dynamic viscosity.

For the dimensionless formulation of (2.1.1), the scales

$$t = \frac{t^* U}{L}, \quad \boldsymbol{x} = \frac{\boldsymbol{x}^*}{L}, \quad \boldsymbol{u}_0 = \frac{\boldsymbol{u}_0^*}{U}, \quad p_0 = \frac{p_0^*}{\rho U^2} \tag{2.1.2}$$

are used, where U and L describe the typical velocity and length scales of the problem to be defined later depending on the system considered. Then the unsteady, incompressible Navier–Stokes and continuity equations can be expressed in non-dimensional form as

$$\partial_t \boldsymbol{u}_0 + \boldsymbol{u}_0 \cdot \nabla \boldsymbol{u}_0 = -\nabla p_0 + \frac{1}{\text{Re}} \nabla^2 \boldsymbol{u}_0, \tag{2.1.3a}$$

$$\nabla \cdot \boldsymbol{u}_0 = 0, \tag{2.1.3b}$$

with the dimensionless Reynolds number

$$\mathrm{Re} = \frac{UL}{\nu}. \tag{2.1.4}$$

In the case of time-independent boundary conditions, a stationary basic flow ($\partial_t \boldsymbol{u}_0 = 0$) exists, which is governed by

$$\boldsymbol{u}_0 \cdot \nabla \boldsymbol{u}_0 = -\nabla p_0 + \frac{1}{\mathrm{Re}} \nabla^2 \boldsymbol{u}_0, \tag{2.1.5a}$$

$$\nabla \cdot \boldsymbol{u}_0 = 0. \tag{2.1.5b}$$

The steady, incompressible Navier–Stokes and continuity equations for a two-dimensional flow in the (x, y) plane can be written by components as

$$\partial_x (u_0 u_0) + \partial_y (v_0 u_0) + \partial_x p_0 - \frac{1}{\mathrm{Re}} (\partial_{xx} u_0 + \partial_{yy} u_0) = 0, \tag{2.1.6a}$$

$$\partial_x (u_0 v_0) + \partial_y (v_0 v_0) + \partial_y p_0 - \frac{1}{\mathrm{Re}} (\partial_{xx} v_0 + \partial_{yy} v_0) = 0, \tag{2.1.6b}$$

$$\partial_x u_0 + \partial_y v_0 = 0. \tag{2.1.6c}$$

In the above equations, the nonlinear convective term of (2.1.5a) $\boldsymbol{u}_0 \cdot \nabla \boldsymbol{u}_0$ has been written in conservative form $\nabla \cdot (\boldsymbol{u}_0 \boldsymbol{u}_0)$, which guarantees conservation of momentum on a discrete mesh (Gresho, 1991).

2.2. Linear Stability Analysis

Considering any initial conditions $\boldsymbol{u}_{\mathrm{tot}}(\boldsymbol{x}, t = 0)$ and $p_{\mathrm{tot}}(\boldsymbol{x}, t = 0)$, which satisfy the same boundary conditions as the basic state, a total flow $\boldsymbol{u}_{\mathrm{tot}}(\boldsymbol{x}, t)$ and $p_{\mathrm{tot}}(\boldsymbol{x}, t)$ exists for $t > 0$, being a solution of

$$\partial_t \boldsymbol{u}_{\mathrm{tot}} + \boldsymbol{u}_{\mathrm{tot}} \cdot \nabla \boldsymbol{u}_{\mathrm{tot}} = -\nabla p_{\mathrm{tot}} + \frac{1}{\mathrm{Re}} \nabla^2 \boldsymbol{u}_{\mathrm{tot}}, \tag{2.2.1a}$$

$$\nabla \cdot \boldsymbol{u}_{\mathrm{tot}} = 0. \tag{2.2.1b}$$

If the difference between the total flow and the basic flow $\tilde{\boldsymbol{u}} = \boldsymbol{u}_{\mathrm{tot}} - \boldsymbol{u}_0$ and $\tilde{p} = p_{\mathrm{tot}} - p_0$ is decaying for $t \to \infty$, the basic flow will be realized. In this case, the basic state is stable, otherwise ($|(\tilde{\boldsymbol{u}}, \tilde{p})| \neq 0$ for $t \to \infty$) unstable, see Chandrasekhar (1961).

Therefore it is interesting to analyse the temporal evolution of small perturbations $|(\tilde{\boldsymbol{u}}, \tilde{p})| \ll 1$ (Drazin & Reid, 1981). Substituting $\boldsymbol{u}_{\text{tot}} = \boldsymbol{u}_0 + \tilde{\boldsymbol{u}}$ and $p_{\text{tot}} = p_0 + \tilde{p}$ into (2.2.1) yields the nonlinear perturbation equations

$$\partial_t \tilde{\boldsymbol{u}} + \boldsymbol{u}_0 \cdot \nabla \tilde{\boldsymbol{u}} + \tilde{\boldsymbol{u}} \cdot \nabla \boldsymbol{u}_0 + \tilde{\boldsymbol{u}} \cdot \nabla \tilde{\boldsymbol{u}} = -\nabla \tilde{p} + \frac{1}{\text{Re}} \nabla^2 \tilde{\boldsymbol{u}}, \qquad (2.2.2\text{a})$$

$$\nabla \cdot \tilde{\boldsymbol{u}} = 0. \qquad (2.2.2\text{b})$$

Owing to the assumption of infinitesimal perturbations, the quadratic term $\tilde{\boldsymbol{u}} \cdot \nabla \tilde{\boldsymbol{u}}$ is neglected in the linear perturbation equations

$$\partial_t \tilde{\boldsymbol{u}} + \boldsymbol{u}_0 \cdot \nabla \tilde{\boldsymbol{u}} + \tilde{\boldsymbol{u}} \cdot \nabla \boldsymbol{u}_0 = -\nabla \tilde{p} + \frac{1}{\text{Re}} \nabla^2 \tilde{\boldsymbol{u}}, \qquad (2.2.3\text{a})$$

$$\nabla \cdot \tilde{\boldsymbol{u}} = 0. \qquad (2.2.3\text{b})$$

The limitation of the above linearization is the fact that the amplitudes of the perturbations are indefinite. Moreover, the analysis is restricted to small disturbances. However, there exist basic states, which are stable with respect to small perturbations, but unstable to larger ones (Criminale *et al.*, 2003).

The computational domain $\mathcal{D}$ (for details confer the sections 4.1.1, 4.2.1 and 4.3.1) is in principle a plane channel with solid walls at the bottom and the top. The inlet boundary is located at the left, and the outlet at the right-hand side of the computational domain $\mathcal{D}$. The three-dimensional perturbation flow $\tilde{\boldsymbol{u}} = (\tilde{u}, \tilde{v}, \tilde{w})^{\text{T}}$ and $\tilde{p}$ satisfy the following boundary conditions

$$\tilde{\boldsymbol{u}} = 0 \quad \text{at the inlet and channel walls}, \qquad (2.2.4\text{a})$$

$$\partial_x \tilde{\boldsymbol{u}} = 0, \tilde{p} = 0 \quad \text{at the outlet channel}, \qquad (2.2.4\text{b})$$

which are the same for all the systems considered in the chapter 4. These inflow and outflow boundary conditions are latest state of the art for the linear instability analysis of open systems (Theofilis, 2011). Note that setting the pressure perturbation to zero at the outlet merely serves to fix the pressure perturbation level to a constant, but does not affect the stability results.

Owing to the infinite extension in the spanwise (z) direction of the systems considered, the general solution of (2.2.3) can be written as a superposition of normal modes

$$\begin{pmatrix} \tilde{\boldsymbol{u}} \\ \tilde{p} \end{pmatrix} (x, y, z, t) = \begin{pmatrix} \hat{\boldsymbol{u}} \\ \hat{p} \end{pmatrix} (x, y) \, \text{e}^{-\gamma t + \text{i} k z} + \text{c.c.}, \qquad (2.2.5)$$

where the complex conjugate (c.c.) renders the perturbations real. Here, $k \in \mathbb{R}$ is a real, positive and continuous wave number in the spanwise direction. The temporal decay rate $\gamma = \sigma + \text{i}\omega \in \mathbb{C}$ of (2.2.5) comprises the real decay rate $\sigma \in \mathbb{R}$ and the oscillation frequency $\omega \in \mathbb{R}$. Here $\hat{\boldsymbol{u}}$ and

$\hat{p}$ represent the complex amplitude/shape functions of the perturbations, which are normalized by setting the maximum norm to 1,

$$\|\hat{\boldsymbol{u}}\|_\infty := \max_{x,y \in \mathcal{D}} \{|\hat{u}|, |\hat{v}|, |\hat{w}|\} = 1 \qquad (2.2.6)$$

over the computational domain $\mathcal{D}$.

Substituting the normal-mode ansatz (2.2.5) into the perturbation equations (2.2.3) yields

$$\begin{aligned}
2\partial_x(u_0\hat{u}) + \partial_y(v_0\hat{u}) + \partial_y(u_0\hat{v}) + \mathrm{i}k(u_0\hat{w}) + \partial_x\hat{p} - \frac{1}{\mathrm{Re}}(\partial_{xx} + \partial_{yy} - k^2)\hat{u} &= \gamma\hat{u} \\
\partial_x(u_0\hat{v}) + 2\partial_y(v_0\hat{v}) + \partial_x(v_0\hat{u}) + \mathrm{i}k(v_0\hat{w}) + \partial_y\hat{p} - \frac{1}{\mathrm{Re}}(\partial_{xx} + \partial_{yy} - k^2)\hat{v} &= \gamma\hat{v} \\
\partial_x(u_0\hat{w}) + \partial_y(v_0\hat{w}) + \mathrm{i}k\hat{p} - \frac{1}{\mathrm{Re}}(\partial_{xx} + \partial_{yy} - k^2)\hat{w} &= \gamma\hat{w} \\
\partial_x\hat{u} + \partial_y\hat{v} + \mathrm{i}k\hat{w} &= 0.
\end{aligned} \qquad (2.2.7)$$

Note that the above equations are already in conservation form by writing $\boldsymbol{u}_0 \cdot \nabla\tilde{\boldsymbol{u}} + \tilde{\boldsymbol{u}} \cdot \nabla\boldsymbol{u}_0$ of (2.2.3a) as $\nabla \cdot (\boldsymbol{u}_0\tilde{\boldsymbol{u}}) + \nabla \cdot (\tilde{\boldsymbol{u}}\boldsymbol{u}_0)$ using $\nabla \cdot \tilde{\boldsymbol{u}} = \nabla \cdot \boldsymbol{u}_0 = 0$.

Equation (2.2.7) constitutes a complex, singular and generalized eigenvalue problem. The eigenvalue $\gamma = \sigma + \mathrm{i}\omega$ represents the temporal decay rate and the corresponding eigenvector $\hat{\boldsymbol{x}} = (\hat{\boldsymbol{u}}, \hat{p})^\mathrm{T}$ describes the amplitude function of the perturbations of (2.2.5). Following the ideas of Theofilis (2003), it is possible to deduce a real eigenvalue problem as the basic flow $\boldsymbol{u}_0 = (u_0, v_0, 0)^\mathrm{T}$ is two-dimensional. Redefining the spanwise component $\hat{w}$ by $\mathrm{i}\breve{w}$ converts the complex eigenvalue problem into a real one

$$\begin{aligned}
2\partial_x(u_0\hat{u}) + \partial_y(v_0\hat{u}) + \partial_y(u_0\hat{v}) - k(u_0\breve{w}) + \partial_x\hat{p} - \frac{1}{\mathrm{Re}}(\partial_{xx} + \partial_{yy} - k^2)\hat{u} &= \gamma\hat{u} \\
\partial_x(u_0\hat{v}) + 2\partial_y(v_0\hat{v}) + \partial_x(v_0\hat{u}) - k(v_0\breve{w}) + \partial_y\hat{p} - \frac{1}{\mathrm{Re}}(\partial_{xx} + \partial_{yy} - k^2)\hat{v} &= \gamma\hat{v} \\
\partial_x(u_0\breve{w}) + \partial_y(v_0\breve{w}) + k\hat{p} - \frac{1}{\mathrm{Re}}(\partial_{xx} + \partial_{yy} - k^2)\breve{w} &= \gamma\breve{w} \\
\partial_x\hat{u} + \partial_y\hat{v} - k\breve{w} &= 0.
\end{aligned} \qquad (2.2.8)$$

With the real eigenvalue problem (2.2.8) only half of the storage is required, which allows flow instabilities to be addressed at substantially higher grid resolutions as in the general complex case (2.2.7).

In the continuous system, infinite eigenmodes $(\tilde{\boldsymbol{u}}, \tilde{p})$ exist, whereas in the numerical discrete case, there are only a finite number of eigenmodes, depending on the dimension of the eigenvalue problem (2.2.8). For $\sigma > 0$ and $\sigma < 0$, the basic flow is linearly stable and unstable, respectively. The neutral Reynolds number Re_n is defined by the fact that $\sigma = 0$ for a certain wave number

k. The critical Reynolds number Re_c, however, is defined by the vanishing of the minimum possible real decay rate σ. Consequently the decay rate σ must be minimized over all discrete eigenvectors $\hat{\boldsymbol{x}} = (\hat{\boldsymbol{u}}, \hat{p})^\mathrm{T}$ and all continuous wave numbers k for a certain Reynolds number Re and for the set of geometric parameters. The definition of the optimization problem is given by

$$\sigma_{\min} = \min_{k \in \mathbb{R}} \sigma\left(k, \mathrm{Re}, \hat{\boldsymbol{x}}\right). \tag{2.2.9}$$

In order to obtain the critical parameters, one has to find that Reynolds and wave numbers, for which (2.2.9) evaluates to zero, i.e. $\sigma_{\min}\left(k = k_c, \mathrm{Re} = \mathrm{Re}_c, \hat{\boldsymbol{x}}\right) = 0$.

2.3. Energy Analysis

For a physical understanding of the instability mechanism, it has proven useful to calculate the kinetic energy transferred between the basic flow and the critical mode (see Albensoeder *et al.*, 2001; Kuhlmann *et al.*, 1997). The spatial distribution of the local energy transfer may provide an insight into the physical instability mechanism. The kinetic energy per unit mass of the perturbation flow is defined as $E_\mathrm{kin} = \int_V \tilde{\boldsymbol{u}}^2 \, \mathrm{d}V / 2$, where the integration is carried out over the volume $V = [0, 2\pi/k] \times \mathcal{D}$ covering a spanwise period $2\pi/k$ of the perturbations times the computational domain $\mathcal{D}$. The rate of change of the kinetic energy $\mathrm{d}E_\mathrm{kin}/\mathrm{d}t$ is governed by the Reynolds–Orr equation (the derivation is given in the appendix A)

$$\frac{1}{D}\frac{\mathrm{d}E_\mathrm{kin}}{\mathrm{d}t} = -1 + \sum_{i=1}^4 \int_V I_i \, \mathrm{d}V - \frac{1}{2}\int_{S_o} I_5 \, \mathrm{d}S = -1 + \sum_{i=1}^4 \int_V I'_i \, \mathrm{d}V - \frac{1}{2}\int_{S_o} I_5 \, \mathrm{d}S, \tag{2.3.1}$$

which is normalized by the total dissipation rate (Johnson, 1998)

$$\begin{aligned} D &= \frac{1}{2\mathrm{Re}} \int_V \left[\nabla \tilde{\boldsymbol{u}} + (\nabla \tilde{\boldsymbol{u}})^\mathrm{T}\right] : \left[\nabla \tilde{\boldsymbol{u}} + (\nabla \tilde{\boldsymbol{u}})^\mathrm{T}\right] \mathrm{d}V \\ &= \frac{1}{\mathrm{Re}} \int_V 2(\partial_x \tilde{u})^2 + 2(\partial_y \tilde{v})^2 + 2(\partial_z \tilde{w})^2 + (\partial_y \tilde{w} + \partial_z \tilde{v})^2 \\ &\quad + (\partial_z \tilde{u} + \partial_x \tilde{w})^2 + (\partial_x \tilde{v} + \partial_y \tilde{u})^2 \, \mathrm{d}V. \end{aligned} \tag{2.3.2}$$

The surface integral

$$-\frac{1}{2}\int_{S_o} I_5 \, \mathrm{d}S = -\frac{1}{2}\int_{S_o} \frac{\tilde{\boldsymbol{u}}^2 u_0}{D} \, \mathrm{d}S \tag{2.3.3}$$

represents the rate of change of the kinetic energy due to convective transport of perturbation energy through the surface of the outlet denoted by S_o, which results from $-\int_V \tilde{\boldsymbol{u}} \cdot (\boldsymbol{u}_0 \cdot \nabla \tilde{\boldsymbol{u}}) \mathrm{d}V$ by integration by parts. Since $\tilde{\boldsymbol{u}} = 0$ at the inlet, no perturbation energy is advected into the system from upstream. Work done by pressure forces $\int_V \tilde{\boldsymbol{u}} \cdot \nabla \tilde{p} \, \mathrm{d}V$ does not arise, since we assume a constant pressure perturbation at the outlet $\tilde{p} = \mathrm{const.}$, see appendix A.

The local energy-transfer rates $-\tilde{\boldsymbol{u}}\cdot(\tilde{\boldsymbol{u}}\cdot\nabla\boldsymbol{u}_0)$ can be decomposed into different terms, depending on the coordinate system, either Cartesian (I_i) or streamline coordinates (I'_i). In Cartesian coordinates, the local energy production rates $-\tilde{\boldsymbol{u}}\cdot(\tilde{\boldsymbol{u}}\cdot\nabla\boldsymbol{u}_0)$ normalized by the total dissipation D read

$$\begin{aligned} I_1 &= -\frac{1}{D}\tilde{u}^2\partial_x u_0, \\ I_2 &= -\frac{1}{D}\tilde{u}\tilde{v}\partial_y u_0, \\ I_3 &= -\frac{1}{D}\tilde{v}\tilde{u}\partial_x v_0, \\ I_4 &= -\frac{1}{D}\tilde{v}^2\partial_y v_0. \end{aligned} \quad (2.3.4)$$

The perturbation field can also be decomposed into components parallel and perpendicular to the basic flow

$$\tilde{\boldsymbol{u}}_\| = \frac{(\tilde{\boldsymbol{u}}\cdot\boldsymbol{u}_0)\boldsymbol{u}_0}{\boldsymbol{u}_0\cdot\boldsymbol{u}_0}, \quad \tilde{\boldsymbol{u}}_\perp = \tilde{\boldsymbol{u}} - \tilde{\boldsymbol{u}}_\|. \quad (2.3.5)$$

In this case the local normalized energy-transfer terms are given by

$$\begin{aligned} I'_1 &= -\frac{1}{D}\tilde{\boldsymbol{u}}_\perp\cdot(\tilde{\boldsymbol{u}}_\perp\cdot\nabla\boldsymbol{u}_0), \\ I'_2 &= -\frac{1}{D}\tilde{\boldsymbol{u}}_\|\cdot(\tilde{\boldsymbol{u}}_\perp\cdot\nabla\boldsymbol{u}_0), \\ I'_3 &= -\frac{1}{D}\tilde{\boldsymbol{u}}_\perp\cdot(\tilde{\boldsymbol{u}}_\|\cdot\nabla\boldsymbol{u}_0), \\ I'_4 &= -\frac{1}{D}\tilde{\boldsymbol{u}}_\|\cdot(\tilde{\boldsymbol{u}}_\|\cdot\nabla\boldsymbol{u}_0). \end{aligned} \quad (2.3.6)$$

The individual terms read as

$$\begin{aligned} I'_1 &= -\frac{1}{D}\left(\tilde{u}_\perp^2\partial_x u_0 + \tilde{u}_\perp\tilde{v}_\perp\partial_y u_0 + \tilde{v}_\perp\tilde{u}_\perp\partial_x v_0 + \tilde{v}_\perp^2\partial_y v_0\right), \\ I'_2 &= -\frac{1}{D}\left(\tilde{u}_\|\tilde{u}_\perp\partial_x u_0 + \tilde{u}_\|\tilde{v}_\perp\partial_y u_0 + \tilde{v}_\|\tilde{u}_\perp\partial_x v_0 + \tilde{v}_\|\tilde{v}_\perp\partial_y v_0\right), \\ I'_3 &= -\frac{1}{D}\left(\tilde{u}_\|\tilde{u}_\perp\partial_x u_0 + \tilde{u}_\perp\tilde{v}_\|\partial_y u_0 + \tilde{v}_\perp\tilde{u}_\|\partial_x v_0 + \tilde{v}_\|\tilde{v}_\perp\partial_y v_0\right), \\ I'_4 &= -\frac{1}{D}\left(\tilde{u}_\|^2\partial_x u_0 + \tilde{u}_\|\tilde{v}_\|\partial_y u_0 + \tilde{v}_\|\tilde{u}_\|\partial_x v_0 + \tilde{v}_\|^2\partial_y v_0\right). \end{aligned} \quad (2.3.7)$$

If $I_i \gtrless 0$ then the local energy transfer associated with the particular term acts as destabilizing or stabilizing. The total local energy production is independent of the decomposition $\sum_{i=1}^4 I_i = \sum_{i=1}^4 I'_i$.

If the rate of change of the kinetic energy of the disturbance flow dE_{kin}/dt is positive, the basic flow is unstable, and vice versa. Hence, the energy analysis can also be used as a consistency check with the linear stability results because on the margin of stability, the rate of change of the kinetic energy has to vanish for the critical mode, i.e.

$$\frac{\mathrm{d}E_{\mathrm{kin}}}{\mathrm{d}t}\Big|_{\sigma=0}= 0. \tag{2.3.8}$$

2.4. Adjoint Analysis

The adjoint of a linear operator is a very important and useful concept in the field of functional analysis. In fluid mechanics, a continuous adjoint formulation has been widely used to tackle problems in receptivity, transition and flow control (Bottaro et al., 2003; Chomaz, 2005; Hill, 1995).

The linear perturbation equations (2.2.3) can be written compactly as $\mathcal{L} \cdot \boldsymbol{q} = 0$ with $\boldsymbol{q} = (\tilde{\boldsymbol{u}}, \tilde{p})^{\mathrm{T}}$. The continuous adjoint operator of $\mathcal{L}$ is denoted by $\mathcal{L}^{\dagger}$ and is defined implicitly by

$$\left\langle \mathcal{L}^{\dagger} \cdot \boldsymbol{q}^{\dagger}, \boldsymbol{q} \right\rangle = \left\langle \boldsymbol{q}^{\dagger}, \mathcal{L} \cdot \boldsymbol{q} \right\rangle, \tag{2.4.1}$$

where the adjoint field is denoted by $\boldsymbol{q}^{\dagger} = (\tilde{\boldsymbol{u}}, \tilde{\mathrm{p}})^{\mathrm{T}}$ and the inner product is given by

$$\left\langle \boldsymbol{q}^{\dagger}, \boldsymbol{q} \right\rangle := \int_{\mathcal{D}} \boldsymbol{q}^{\dagger} \cdot \boldsymbol{q}\, \mathrm{d}\mathcal{D}. \tag{2.4.2}$$

Integration by parts yields the adjoint linear perturbation equations (Blackburn et al., 2008; Cantwell et al., 2010)

$$-\partial_t \tilde{\boldsymbol{u}} - \boldsymbol{u}_0 \cdot \nabla \tilde{\boldsymbol{u}} + \tilde{\boldsymbol{u}} \cdot (\nabla \boldsymbol{u}_0)^{\mathrm{T}} = -\nabla \tilde{\mathrm{p}} + \frac{1}{\mathrm{Re}} \nabla^2 \tilde{\boldsymbol{u}}, \tag{2.4.3a}$$

$$\nabla \cdot \tilde{\boldsymbol{u}} = 0, \tag{2.4.3b}$$

with the adjoint normal mode ansatz

$$\begin{pmatrix} \tilde{\boldsymbol{u}} \\ \tilde{\mathrm{p}} \end{pmatrix}(x, y, z, t) = \begin{pmatrix} \hat{\boldsymbol{u}} \\ \hat{\mathrm{p}} \end{pmatrix}(x, y)\, \mathrm{e}^{+\bar{\gamma}t + \mathrm{i}kz} + \mathrm{c.c.} \tag{2.4.4}$$

As the adjoint system is only well posed in the negative time direction, it is sometimes referred to as the backward system.

The eigenvalue problem (2.2.8) can be written compactly as $\boldsymbol{A} \cdot \hat{\boldsymbol{x}} = \gamma \boldsymbol{B} \cdot \hat{\boldsymbol{x}}$ with $\hat{\boldsymbol{x}} = (\hat{u}, \hat{v}, \hat{w}, \hat{p})^{\mathrm{T}}$. Since the operators $\boldsymbol{A}$ and $\boldsymbol{B}$ are real and linear, the adjoint modes can be easily computed by taking the transpose of these operators

$$\boldsymbol{A}^{\mathrm{T}} \cdot \boldsymbol{\mathcal{X}} = \bar{\gamma}\, \boldsymbol{B}^{\mathrm{T}} \cdot \boldsymbol{\mathcal{X}} \tag{2.4.5}$$

with $\boldsymbol{\mathcal{X}} = (\hat{u}, \hat{v}, \hat{w}, \hat{p})^T$, where the correct boundary conditions for the adjoint fields are automatically taken into account. The direct and adjoint eigenvectors are normalized by imposing the conditions

$$\|\hat{\boldsymbol{u}}\| := \sqrt{\langle \hat{\boldsymbol{u}}, \hat{\boldsymbol{u}} \rangle}, \quad \int_{\mathcal{D}} \hat{\boldsymbol{u}} \cdot \hat{\boldsymbol{u}} \, d\mathcal{D} = 1 \quad (2.4.6a)$$

$$\langle \hat{\boldsymbol{u}}, \check{\boldsymbol{u}} \rangle = \int_{\mathcal{D}} \hat{\boldsymbol{u}} \cdot \check{\boldsymbol{u}} \, d\mathcal{D} = 1 \quad (2.4.6b)$$

over the computational domain $\mathcal{D}$. Note that the adjoint eigenvector is not scaled according to the naturally defined norm based upon the inner product of the space itself $\|\check{\boldsymbol{u}}\|$, because the normalization (2.4.6b) simplifies (2.4.8).

The adjoint velocity fields $|\check{\boldsymbol{u}}| = |(\check{u}, \check{v}, \check{w})^T|$ determine the flow regions, which are most receptive to initial conditions and to momentum forcing, whereas $|\check{p}|$ describes the receptivity to mass injection. Moreover, an adjoint analysis allows us to identify the sensitivity of the eigenvalue spectrum of the original problem to spatially localized perturbations. Giannetti & Luchini (2007) have shown that the drift $\Delta\gamma$ due to a perturbation localized at (x_0, y_0) of the eigenvalue γ associated with the eigenvector $\hat{\boldsymbol{u}}$ is bounded by

$$|\Delta\gamma| \leq c_0 \Upsilon(x_0, y_0) \quad (2.4.7)$$

with

$$\Upsilon(x, y) = \frac{|\check{\boldsymbol{u}}(x, y)| \, |\hat{\boldsymbol{u}}(x, y)|}{\underbrace{\langle \hat{\boldsymbol{u}}, \check{\boldsymbol{u}} \rangle}_{=1}}, \quad (2.4.8)$$

where $c_0 > 0$ of (2.4.7) is a measure for the strength of the localized forcing. Hence, the scaled product $\Upsilon(x, y)$ between the direct and adjoint velocity eigenmodes represents a measure for the sensitivity of the temporal growth rates to local perturbations.

3. Numerical Implementation

3.1. Newton's Method

The nonlinear system of equations (2.1.5) for the steady two-dimensional basic state is solved with Newton's method, also known as Newton–Raphson method. This scheme converges, in the ideal case, quadratically to the final solution (Kelley, 2003). Newton's method is given by

$$\boldsymbol{J}(\boldsymbol{x}^n) \cdot \delta\boldsymbol{x} = -\boldsymbol{f}(\boldsymbol{x}^n) \qquad (3.1.1\text{a})$$

$$\boldsymbol{x}^{n+1} = \boldsymbol{x}^n + \delta\boldsymbol{x} \qquad (3.1.1\text{b})$$

with $\boldsymbol{x} = (u_0, v_0, p_0)^\mathrm{T}$. $\boldsymbol{J}(\boldsymbol{x}^n)$ represents the Jacobian evaluated at the actual solution vector $\boldsymbol{x}^n$ and $\boldsymbol{f}(\boldsymbol{x}^n)$ is a vector-valued function, representing the nonlinear residual of (2.1.5). Inserting (3.1.1b) into (2.1.5) yields

$$\boldsymbol{u}_0^n \cdot \nabla \delta\boldsymbol{u} + \delta\boldsymbol{u} \cdot \nabla \boldsymbol{u}_0^n + \boldsymbol{u}_0^n \cdot \nabla \boldsymbol{u}_0^n + \delta\boldsymbol{u} \cdot \nabla \delta\boldsymbol{u} + \nabla(p_0^n + \delta p) - \frac{1}{\mathrm{Re}}\nabla^2(\boldsymbol{u}_0^n + \delta\boldsymbol{u}) = 0$$
$$\nabla \cdot (\boldsymbol{u}_0^n + \delta\boldsymbol{u}) = 0. \qquad (3.1.2)$$

With Newton's linearization the quadratic term $\delta\boldsymbol{u} \cdot \nabla \delta\boldsymbol{u}$ is neglected, see ur Rehman *et al.* (2006). All terms being linear in δ of (3.1.2) are moved to the left-hand side, leading to

$$\boldsymbol{u}_0^n \cdot \nabla \delta\boldsymbol{u} + \delta\boldsymbol{u} \cdot \nabla \boldsymbol{u}_0^n + \nabla \delta p - \frac{1}{\mathrm{Re}}\nabla^2 \delta\boldsymbol{u} = -\left(\boldsymbol{u}_0^n \cdot \nabla \boldsymbol{u}_0^n + \nabla p_0^n - \frac{1}{\mathrm{Re}}\nabla^2 \boldsymbol{u}_0^n\right)$$
$$\nabla \cdot \delta\boldsymbol{u} = -(\nabla \cdot \boldsymbol{u}_0^n). \qquad (3.1.3)$$

The above equation represents the linear system (3.1.1a), which is solved for the unknowns $\delta\boldsymbol{x} = (\delta\boldsymbol{u}, \delta p)^\mathrm{T}$.

If Newton's iteration converges, i.e. $\boldsymbol{x}^n \to \boldsymbol{x}$, the correction vector $\delta\boldsymbol{x} = 0$ and the right-hand side of (3.1.3) evaluates to zero ($-\boldsymbol{f}(\boldsymbol{x}) = 0$). The Jacobian $\boldsymbol{J}(\boldsymbol{x})$ of the stationary Navier–Stokes equations can be readily obtained by applying Newton's linearization on the convective terms, yielding

$$\nabla \cdot (\boldsymbol{u}_0 \boldsymbol{u}_0) \approx \nabla \cdot (\bar{\boldsymbol{u}}_0 \boldsymbol{u}_0) + \nabla \cdot (\boldsymbol{u}_0 \bar{\boldsymbol{u}}_0). \qquad (3.1.4)$$

Here linearization takes place around $\bar{\boldsymbol{u}}_0$, representing the solution vector from the previous iteration step. Note the similarities to the linear perturbation equations (2.2.3).

The major drawback of Newton's method is its local convergence, which means that the scheme only converges, if the initial guess does not differ too much from the final solution. In order to obtain a good initial guess, a few Picard iteration steps are conducted. Within Picard's linearization, the velocity vector of the previous iteration step $\bar{\boldsymbol{u}}_0$ is substituted in the convective term

$$\nabla \cdot (\boldsymbol{u}_0 \boldsymbol{u}_0) \approx \nabla \cdot (\bar{\boldsymbol{u}}_0 \boldsymbol{u}_0), \tag{3.1.5}$$

leading to

$$\partial_x(\bar{u}_0 u_0) + \partial_y(\bar{v}_0 u_0) + \partial_x p_0 - \frac{1}{\mathrm{Re}}(\partial_{xx} u_0 + \partial_{yy} u_0) = 0,$$
$$\partial_x(\bar{u}_0 v_0) + \partial_y(\bar{v}_0 v_0) + \partial_y p_0 - \frac{1}{\mathrm{Re}}(\partial_{xx} v_0 + \partial_{yy} v_0) = 0, \tag{3.1.6}$$
$$\partial_x u_0 + \partial_y v_0 = 0.$$

Picard's method is terminated if either 5 iteration steps are reached or alternatively the residual is below 10^{-2}. The residual at the nth iteration step is defined as

$$\mathrm{res}(\boldsymbol{x}^n) = \max\left\{ \frac{|\boldsymbol{f}(\boldsymbol{x}^n)|}{|\boldsymbol{x}^n|}, \frac{|\boldsymbol{f}(\boldsymbol{x}^n)|}{\sqrt{N}} \right\}, \tag{3.1.7}$$

where N is the number of unknowns. For reasons of safety, the residual is the maximum between the scaled Euclidean and the root-mean-square (RMS) norm, which is defined by $|\boldsymbol{f}(\boldsymbol{x}^n)|/\sqrt{N}$. Taking simply the Euclidean norm $|\boldsymbol{f}(\boldsymbol{x})|$ without scaling is not a good choice as it depends on the number of unknowns N.

Moreover, a polynomial line search strategy for globalizing Newton's method is incorporated, following the ideas of Kelley (1995). In the line search approach, the direction of the correction vector $\delta \boldsymbol{x}$ of (3.1.1) is used, but the step size is reduced, if necessary. This globalization strategy is implemented between (3.1.1a) and (3.1.1b). It is activated, if the new residual is larger than the previous one, namely

$$\mathrm{res}(\boldsymbol{x}^n + \underbrace{\beta}_{=1} \delta \boldsymbol{x}) > \mathrm{res}(\boldsymbol{x}^n). \tag{3.1.8}$$

If the above relation is satisfied, 10 discrete values of β are taken in the interval $\beta_i \in \,]0.1, 1[$ and the residuals are computed. Then a parabola is fitted to the 3 residuals for the values β_{i-1}, β_{i+1} and β_i, at which the residual takes its minimum within the 10 discrete values. The parabola gives a better estimate of the optimum value of β_{opt}, which is then used in the actual iteration for reducing the step size in the correction vector

$$\boldsymbol{x}^{n+1} = \boldsymbol{x}^n + \beta_{\mathrm{opt}} \delta \boldsymbol{x}. \tag{3.1.9}$$

Finally, Newton's method is terminated when the residual (3.1.7) is below the square root of the machine accuracy $\sqrt{\epsilon_{\text{mach}}} \approx 10^{-8}$ or a similar accuracy level.

In addition to some Picard iterations and the polynomial line search strategy, a natural continuation method is implemented, providing good initial guesses for the next parameter combination by following the solution path (Howell, 2009). In simple or classical continuation, the converged solution of $f(x) =: f(x(\text{Re}_{\text{old}})) = 0$ serves as an initial guess for the solution to be computed with a different Reynolds number, i.e. $x^0(\text{Re}_{\text{new}}) = x(\text{Re}_{\text{old}})$. This approach may be inefficient for small slopes $\partial x(\text{Re}_{\text{old}})/\partial \text{Re}_{\text{old}}$ on the solution branch, where there is only a moderate change in x for a significant range of Re. Simple continuation may encounter difficulties if the solution manifold in the (x, Re)-space exhibits large changes in x for only moderate changes in Re (high slopes). The method can be improved by incorporating the slope $\partial x(\text{Re}_{\text{old}})/\partial \text{Re}_{\text{old}}$ of the solution path at $x(\text{Re}_{\text{old}})$, known as natural or tangent continuation, see Howell (2009). Hereby, the initial guess is formed by the equation of a tangent

$$x^0(\text{Re}_{\text{new}}) = x(\text{Re}_{\text{old}}) + |\text{Re}_{\text{new}} - \text{Re}_{\text{old}}| \frac{\partial x(\text{Re}_{\text{old}})}{\partial \text{Re}_{\text{old}}}. \quad (3.1.10)$$

In contrast to simple continuation, natural continuation represents a method of second order. In Haselgrove (1961) it is also shown that (3.1.10) may be improved to third order accuracy by incorporating two tangential gradients, which requires, however, the storage of two previous solution vectors.

The linear system of equations (3.1.1) is solved directly with an efficient solver for sparse matrices, provided by MATLAB's backslash ('\') operator. A different approach, called the Jacobian-free Newton–Krylov method, is presented in the appendix B, which, however, did not turn out to be superior, at least for 2D and 3D problems, where one direction is treated analytically as in the present stability analysis.

In numerical algebra it proved to be very useful to reorder sparse matrices prior to LU or Cholesky factorizations, which are performed by MATLAB's backslash ('\') and eigs (eigensolver) operators, depending on the input matrix. Two widely used reordering algorithms can be found in the literature, namely the approximate minimum degree (AMD) and the reverse Cuthill-McKee (RCM) permutations, see Amestoy et al. (1996) and Gilbert et al. (1992), respectively. While AMD produces a structure with large blocks of connected zeros, RCM reduces the bandwidth of the resulting matrix. MATLAB's implementation of the column AMD `colamd` is slightly faster than the symmetric RCM `symrcm`. Moreover, it takes slightly less time to solve the linear system and to compute the eigenvalues with `colamd` than with `symrcm`.

Figure 3.1 shows the matrices of coefficients of the linearized Navier–Stokes equations (2.1.6) with Picard's (3.1.5) and Newton's (3.1.4) linearization. The system matrix 3.1(b) is then reordered with the `symrcm` and `colamd` algorithm, respectively.

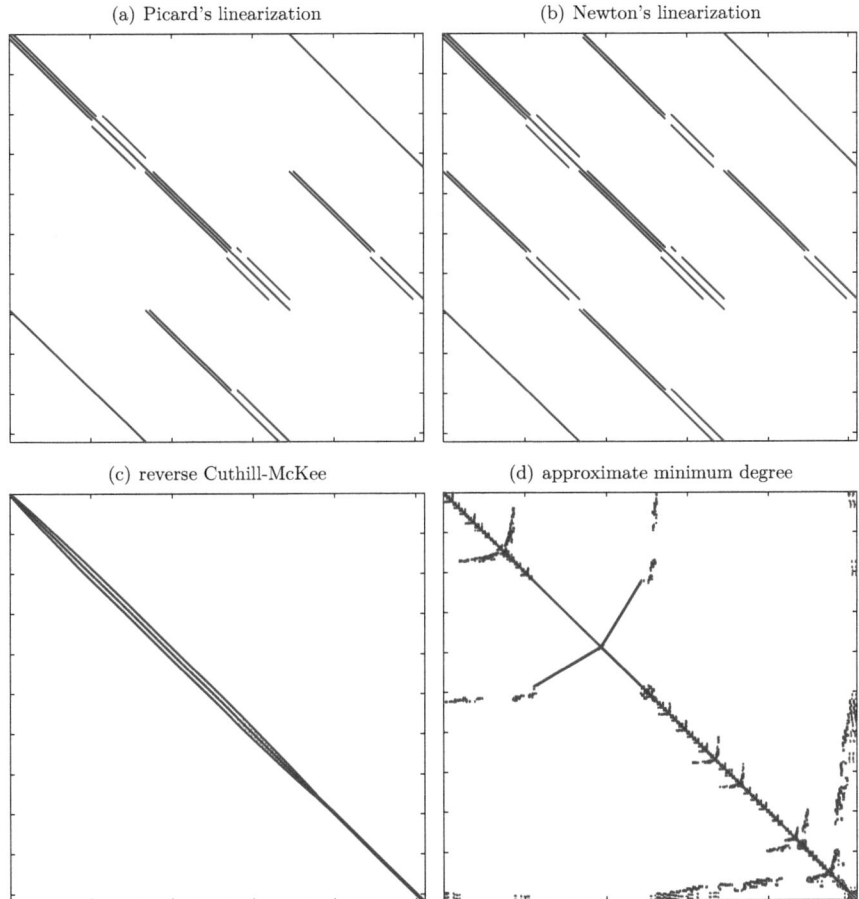

Figure 3.1.: Sparsity pattern of the matrices of coefficients of the linearized Navier–Stokes equations (2.1.6) ((a) and (b)). The RCM and AMD reordering schemes were applied to the matrix (b).

3.2. Finite-Volume Discretization

The conservative form of the Navier–Stokes equations, guaranteeing conservation of momentum on a discrete mesh (Gresho, 1991), is discretized by finite volumes on a staggered rectangular grid using primitive variables. Derivatives and intermediate values are approximated by central differences and linear interpolation, leading to a method of second order. The advantage of a finite-volume approach is that it works well at singularities from, e.g., sharp corners of the geometry.

Before discretization, the nonlinear term of (2.1.5) is linearized using Newton's approach. Inserting (3.1.4) into (2.1.5) gives

$$2\partial_x(\bar{u}_0 u_0) + \partial_y(\bar{v}_0 u_0) + \partial_y(\bar{u}_0 v_0) + \partial_x p_0 - \frac{1}{\text{Re}}(\partial_{xx} u_0 + \partial_{yy} u_0) = 0, \quad (3.2.1\text{a})$$

$$\partial_x(\bar{v}_0 u_0) + \partial_x(\bar{u}_0 v_0) + 2\partial_y(\bar{v}_0 v_0) + \partial_y p_0 - \frac{1}{\text{Re}}(\partial_{xx} v_0 + \partial_{yy} v_0) = 0, \quad (3.2.1\text{b})$$

$$\partial_x u_0 + \partial_y v_0 = 0. \quad (3.2.1\text{c})$$

Note the differences (in blue colour) to the equations (3.1.6), which were obtained using Picard's linearization. Equation (3.2.1) can be expressed in compact form (see (3.1.1))

$$\boldsymbol{J}(\tilde{\boldsymbol{x}}) \cdot \boldsymbol{x} = \underbrace{\boldsymbol{J}(\tilde{\boldsymbol{x}}) \cdot \tilde{\boldsymbol{x}} - \boldsymbol{f}(\tilde{\boldsymbol{x}})}_{=0} = 0. \quad (3.2.2)$$

Spatial discretization will be carried out on a staggered grid, which remains the method of choice for orthogonal grids, as the pressure gradient can be formed without interpolation (Wesseling, 2001). Within a staggered grid, the pressure resides at the cell centres and the normal velocity components are placed at the cell faces, thus representing a combination of vertex-centred and cell-centred discretization. The control volume with centre at $x_{j,k}$ is denoted $\Omega_{j,k} = h^x_{j,k} h^y_{j,k}$ with horizontal and vertical lengths $h^x_{j,k}$ and $h^y_{j,k}$, respectively. Thus there are $(J+1) \times K$ u-cells, $J \times (K+1)$ v and $J \times K$ p-nodes.

The continuity equation is discretized by integrating over $\Omega_{j,k}$, resulting in

$$\int_{\Omega_{j,k}} \nabla \cdot \boldsymbol{u} \, d\Omega \cong h^y_{j,k} u\big|^{j+1/2,k}_{j-1/2,k} + h^x_{j,k} v\big|^{j,k+1/2}_{j,k-1/2} = 0$$

$$\Rightarrow \frac{1}{h^x_{j,k}} u\big|^{j+1/2,k}_{j-1/2,k} + \frac{1}{h^y_{j,k}} v\big|^{j,k+1/2}_{j,k-1/2} = 0. \quad (3.2.3)$$

Note that for convenience of notation the subscript 0 in the flow quantities has been dropped. In the prevailing subsection, a subscript will denote a derivative, e.g. $\partial_x u =: u_x$.

The control volume for the u-component of (3.2.1a) consists of the union of half of $\Omega_{j,k}$ and half of $\Omega_{j+1,k}$. It will be denoted as $\Omega_{j+1/2,k}$ with a length in the x direction of $h^x_{j+1/2,k}$. Finite volume integration over the u-component of (3.2.1a) gives

$$\int_{\Omega_{j+1/2,k}} \left[(2\bar{u}u + p - \frac{1}{\text{Re}}u_x)_x + (\bar{v}u + \bar{u}v - \frac{1}{\text{Re}}u_y)_y \right] d\Omega \cong$$
$$\cong h^y_{j,k}(2\bar{u}u + p - \frac{1}{\text{Re}}u_x)^{j+1,k}_{j,k} + h^x_{j+1/2,k}(\bar{v}u + \bar{u}v - \frac{1}{\text{Re}}u_y)^{j+1/2,k+1/2}_{j+1/2,k-1/2} = 0 \qquad (3.2.4)$$
$$\Rightarrow \frac{1}{h^x_{j+1/2,k}}(2\bar{u}u + p - \frac{1}{\text{Re}}u_x)^{j+1,k}_{j,k} + \frac{1}{h^y_{j,k}}(\bar{v}u + \bar{u}v - \frac{1}{\text{Re}}u_y)^{j+1/2,k+1/2}_{j+1/2,k-1/2} = 0.$$

Integration over the v-component of (3.2.1b) yields

$$\int_{\Omega_{j,k+1/2}} \left[(\bar{v}u + \bar{u}v - \frac{1}{\text{Re}}v_x)_x + (2\bar{v}v + p - \frac{1}{\text{Re}}v_y)_y \right] d\Omega \cong$$
$$\cong h^y_{j,k+1/2}(\bar{v}u + \bar{u}v - \frac{1}{\text{Re}}v_x)^{j+1/2,k+1/2}_{j-1/2,k+1/2} + h^x_{j,k}(2\bar{v}v + p - \frac{1}{\text{Re}}v_y)^{j,k+1}_{j,k} = 0 \qquad (3.2.5)$$
$$\Rightarrow \frac{1}{h^x_{j,k}}(\bar{v}u + \bar{u}v - \frac{1}{\text{Re}}v_x)^{j+1/2,k+1/2}_{j-1/2,k+1/2} + \frac{1}{h^y_{j,k+1/2}}(2\bar{v}v + p - \frac{1}{\text{Re}}v_y)^{j,k+1}_{j,k} = 0.$$

Derivatives are approximated by a second order central difference scheme

$$\begin{aligned}
u_x|_{j,k} &\cong (u_{j+1/2,k} - u_{j-1/2,k})/h^x_{j,k}, \\
u_y|_{j+1/2,k-1/2} &\cong (u_{j+1/2,k} - u_{j+1/2,k-1})/\left[(h^y_{j+1/2,k} + h^y_{j+1/2,k-1})/2\right], \\
v_x|_{j-1/2,k+1/2} &\cong (v_{j,k+1/2} - v_{j-1,k+1/2})/\left[(h^x_{j,k+1/2} + h^x_{j-1,k+1/2})/2\right], \\
v_y|_{j,k} &\cong (v_{j,k+1/2} - v_{j,k-1/2})/h^y_{j,k}.
\end{aligned} \qquad (3.2.6)$$

For the inertia terms no derivatives are required. But the unknowns must be evaluated at points, which do not occur on the grid (inter-grid points). Linear interpolation, which is referred to as the central scheme, is also of second order and results in

$$
\begin{aligned}
(\bar{u}u)_{j,k} &\cong \frac{\bar{u}_{j+1/2,k}h^x_{j-1/2,k} + \bar{u}_{j-1/2,k}h^x_{j+1/2,k}}{h^x_{j-1/2,k} + h^x_{j+1/2,k}} \frac{u_{j+1/2,k}h^x_{j-1/2,k} + u_{j-1/2,k}h^x_{j+1/2,k}}{h^x_{j-1/2,k} + h^x_{j+1/2,k}}, \\
(\bar{v}u)_{j+1/2,k-1/2} &\cong \frac{\bar{v}_{j+1,k-1/2}h^x_{j,k-1/2} + \bar{v}_{j,k-1/2}h^x_{j+1,k-1/2}}{h^x_{j+1,k-1/2} + h^x_{j,k-1/2}} \frac{u_{j+1/2,k-1}h^y_{j+1/2,k} + u_{j+1/2,k}h^y_{j+1/2,k-1}}{h^y_{j+1/2,k} + h^y_{j+1/2,k-1}}, \\
(\bar{u}v)_{j+1/2,k-1/2} &\cong \frac{\bar{u}_{j+1/2,k-1}h^y_{j+1/2,k} + \bar{u}_{j+1/2,k}h^y_{j+1/2,k-1}}{h^y_{j+1/2,k} + h^y_{j+1/2,k-1}} \frac{v_{j+1,k-1/2}h^x_{j,k-1/2} + v_{j,k-1/2}h^x_{j+1,k-1/2}}{h^x_{j+1,k-1/2} + h^x_{j,k-1/2}}, \\
(\bar{v}u)_{j-1/2,k+1/2} &\cong \frac{\bar{v}_{j-1,k+1/2}h^x_{j,k+1/2} + \bar{v}_{j,k+1/2}h^x_{j-1,k+1/2}}{h^x_{j-1,k+1/2} + h^x_{j,k+1/2}} \frac{u_{j-1/2,k+1}h^y_{j-1/2,k} + u_{j-1/2,k}h^y_{j-1/2,k+1}}{h^y_{j-1/2,k+1} + h^y_{j-1/2,k}}, \\
(\bar{u}v)_{j-1/2,k+1/2} &\cong \frac{\bar{u}_{j-1/2,k+1}h^y_{j-1/2,k} + \bar{u}_{j-1/2,k}h^y_{j-1/2,k+1}}{h^y_{j-1/2,k+1} + h^y_{j-1/2,k}} \frac{v_{j-1,k+1/2}h^x_{j,k+1/2} + v_{j,k+1/2}h^x_{j-1,k+1/2}}{h^x_{j-1,k+1/2} + h^x_{j,k+1/2}}, \\
(\bar{v}v)_{j,k} &\cong \frac{\bar{v}_{j,k+1/2}h^y_{j,k-1/2} + \bar{v}_{j,k-1/2}h^y_{j,k+1/2}}{h^y_{j,k-1/2} + h^y_{j,k+1/2}} \frac{v_{j,k+1/2}h^y_{j,k-1/2} + v_{j,k-1/2}h^y_{j,k+1/2}}{h^y_{j,k-1/2} + h^y_{j,k+1/2}}.
\end{aligned}
$$
(3.2.7)

Since no interpolation is needed for the pressure p on a staggered grid, spurious modes unlike in the case of a collocated grid (at least for second order discretization) will not arise (see Ferziger & Perić, 2002).

Equations (3.2.4), (3.2.5), (3.2.6) and (3.2.7) are valid for the interior control volumes. The equations for the boundary cells have to be adjusted according to the prescribed boundary conditions. If a cell occurs outside of the domain (ghost variable), it will be eliminated by taking into account the relevant boundary condition. Therefore no explicit ghost variables are needed in the present implementation. Thereby it is assumed that the outer cell has the same size as the adjacent one inside the domain. Thus taking the arithmetic mean for the central scheme/linear interpolation is justified at the boundaries, even on a non-uniform grid.

The geometries considered in the chapter 4 can be modelled everywhere with Dirichlet boundary conditions (prescribed velocities), except for the outlet, where a homogeneous Neumann condition, i.e $\boldsymbol{u}_x = 0$, is prescribed. This boundary condition is recommended by Versteeg & Malalasekera (2007) in order to avoid spurious numerical oscillations at the outlet. $\boldsymbol{u}_x = 0$ is also physically justified by assuming that the base flow is fully developed at the outlet. Owing to the staggered grid, the Dirichlet boundary conditions are easy to implement for the variables, which are located exactly at the physical boundaries, namely for the u-component in the x direction and for the v-component in the y direction. For the u-equation in the viscous term, a derivative in the y direction u_y is needed. For the bottom-wall ($y = 0$)

cell, for instance, $u_y|_{j+1/2,1/2} \cong (u_{j+1/2,1} - u_{j+1/2,0})/h^y_{j+1/2,k}$ is needed. $u_{j+1/2,0}$ is located outside of the domain, but can be eliminated by the relation

$$\frac{u_{j+1/2,1} + u_{j+1/2,0}}{2} = 0, \quad (3.2.8)$$

where the right-hand side of (3.2.8) represents the wall velocity. With the above relation, one obtains $u_y|_{j+1/2,1/2} \cong 2u_{j+1/2,1}/h^y_{j+1/2,k}$. At the outlet, for $v_x|^{J+1/2,k+1/2} \cong (v^{J+1,k+1/2} - v^{J,k+1/2})/h^x_{J,k+1/2}$ the ghost variable $v_{J+1,k+1/2}$ occurs, which can be eliminated by the homogeneous Neumann condition

$$\frac{v^{J+1,k+1/2} - v^{J,k+1/2}}{h^x_{J,k+1/2}} = 0, \quad (3.2.9)$$

yielding $v_x|^{J+1/2,k+1/2} \cong (v^{J+1,k+1/2} - v^{J+1,k+1/2})/h^x_{J,k+1/2} = 0$. In (3.2.4) the pressure is needed at the outlet at the position $p^{J+1,K}$. Likewise an explicit ghost cell is not required due to

$$\frac{p^{J+1,K} + p^{J,K}}{2} = p_\infty, \quad (3.2.10)$$

where p_∞ represents the ambient pressure at the outlet. Owing to the staggered grid, the pressure equations for the cells boundaries do not have to be adjusted at Dirichlet boundary conditions.

Once the basic state is obtained, it is inserted into the linear perturbation equations (2.2.8), which are discretized on the same grid. Finite volumes are used in the (x,y) plane, where the third component of the perturbation field $\breve{w}$ resides with the pressure nodes.

3.3. Grid Generation

While the truncation error of the central-difference quotient is of second order on an equidistant grid, it decreases formally to first order on an inhomogeneous computational domain. The discretization error of, for instance, the first derivative with central differencing is defined by the leading term of the Taylor series expansion around the point x_i

$$\phi_{i+1} - \phi_{i-1} = (\Delta x_{i+1} + \Delta x_i)\phi_x|_i + \frac{\Delta x_{i+1}^2 - \Delta x_i^2}{2}\phi_{xx}|_i + \frac{\Delta x_{i+1}^3 + \Delta x_i^3}{3!}\phi_{xxx}|_i + \text{h.o.t.}, \quad (3.3.1)$$

which results in

$$\phi_x|_i = \frac{\phi_{i+1} - \phi_{i-1}}{\Delta x_{i+1} + \Delta x_i} - \frac{\Delta x_{i+1} - \Delta x_i}{2}\phi_{xx}|_i - \frac{\Delta x_{i+1}^3 + \Delta x_i^3}{3!(\Delta x_{i+1} + \Delta x_i)}\phi_{xxx}|_i + \text{h.o.t.}. \quad (3.3.2)$$

Therefore the central difference quotient is of second order only on a homogeneous grid $\Delta x_{i+1} = \Delta x_i$. If, however, an inhomogeneous grid is refined systematically, the discretization error converges asymptotically to second order (Ferziger & Perić, 2002). To minimize the first-order error, the size of neighbouring cells should not differ too much. The leading-order error can be written as

$$\frac{\Delta x_{i+1} - \Delta x_i}{2}\phi_{xx}|_i = \frac{\Delta x_i(\delta - 1)}{2}\phi_{xx}|_i, \qquad (3.3.3)$$

with the stretching factor $\delta = \Delta x_{i+1}/\Delta x_i$. According to Fletcher (1988) the grid stretching factor should be within $0.8 < \delta < 1.2$, whereby δ depends on the number of cells and the gradients to be resolved. In times of high computing power, the stretching factors are getting smaller and I personally used $0.95 \leq \delta \leq 1.05$.

For distributing the grid points along a segment line with length Δs, several one-dimensional distribution function exist in the literature. The geometrical stretching function is defined as

$$s = \Delta s_1 \frac{\delta^n - 1}{\delta - 1}, \qquad (3.3.4)$$

where s represents the position of the nodes and Δs_1 the size of the first cell. For $n = 0 \ldots N$, the nodes are put on the edges of the segment line, which is needed for u in the x direction and for v in the y direction on a staggered grid. For the interior points, such as the pressure nodes, $n = (0 \ldots N - 1) + 0.5$.

Equation (3.3.4) is getting singular, if $\delta = 1$. Setting $\varepsilon = |\delta - 1|$ and expanding s in a Taylor series for $\varepsilon \ll 1$ gives

$$s = \Delta s_1 N \left[1 + \frac{N-1}{2}(\delta - 1) + \frac{1}{6}(N-1)(N-2)(\delta - 1)^2 + O(\delta - 1)^3\right]. \qquad (3.3.5)$$

Another very popular distribution function is the hyperbolic tangent stretching function, which dates back to Vinokur (1983). The one-sided refinement is defined as

$$s = \Delta s \left[1 + \frac{\tanh\left(b/2\left\{n/N - 1\right\}\right)}{\tanh\left(b/2\right)}\right]. \qquad (3.3.6)$$

The parameter b determines the maximum allowable rate of change of adjacent cells in percent (i.e. $b \leq 5$) and can either be specified or determined by solving the transcendental equation

$$\frac{\sinh b}{b} - \frac{\Delta s}{N \Delta s_1} = 0. \qquad (3.3.7)$$

For a double-sided refinement the grid spacing on the left Δs_1 and on the right side Δs_2 have to be specified. Now the hyperbolic tangent stretching function is defined as

$$s_0 = \frac{1}{2}\left[1 + \frac{\tanh\left(b\{n/N - 0.5\}\right)}{\tanh(b/2)}\right]$$
$$s = \Delta s \frac{s_0}{\sqrt{\Delta s_2/\Delta s_1} + (1 - \sqrt{\Delta s_2/\Delta s_1})s_0}. \tag{3.3.8}$$

b can be computed from the solution of

$$\frac{\sinh b}{b} - \frac{\Delta s}{N\sqrt{\Delta s_2 \Delta s_1}} = 0, \tag{3.3.9}$$

where the following condition has to be fulfilled

$$N < \frac{\sqrt{\Delta s(1 - 10^{-6})}}{\sqrt{\Delta s_2 \Delta s_1}}. \tag{3.3.10}$$

In Thompson et al. (1985) an error analysis of several distribution functions (exponential, hyperbolic sine, error function,...) can be found and the hyperbolic tangent performed best in the global sense. The geometrical stretching function (3.3.4) was not covered by this analysis. The main difference between the hyperbolic tangent and the geometrical distribution function is the fact that in the latter case the rate of change among adjacent cells is constant, whereas it changes smoothly for the hyperbolic tangent refinement. Moreover, if one applies grid refinement on, for instance, the outflow boundary, the rate of change among adjacent cells is biggest near the boundary and decreases smoothly for the next cells for the hyperbolic tangent distribution function (3.3.6). Therefore, the geometrical stretching function (3.3.4) with $\delta = 1.03$ was used for the systems addressed in the chapter 4. As a rule of thumb, approximately the same number of grid points are obtained, if the stretching factor is set to $\delta = 1.03$ in (3.3.4) and $b = 5$ in (3.3.6).

3.4. Eigenvalue-Detection Strategies

Equation (2.2.8) constitutes a real, singular and generalized eigenvalue problem

$$\boldsymbol{A} \cdot \hat{\boldsymbol{x}} = \gamma \boldsymbol{B} \cdot \hat{\boldsymbol{x}}, \tag{3.4.1}$$

where $\hat{\boldsymbol{x}} = (\hat{u}, \hat{v}, \hat{w}, \hat{p})^{\mathrm{T}}$ and $\boldsymbol{A}$ and $\boldsymbol{B}$ are linear and real operators. Computing the whole spectrum of (3.4.1) is infeasible because the size of the problem is too large in real applications. Computing only selected eigenvalues by an iterative eigensolver is challenging, because $\boldsymbol{A}$ is nonsymmetric and, more important, $\boldsymbol{B}$ is singular. Owing to this singularity, (3.4.1) has infinite eigenvalues, which are physically irrelevant and cause numerical difficulties.

The eigenvalue problem is solved with an implicitly restarted Arnoldi algorithm as provided in the ARPACK software library and MATLAB's `eigs` command (Lehoucq & Sorensen, 1996). Because of the normal mode ansatz (2.2.5), the eigenvalues with smallest real part are of interest. However, they cannot be computed directly with Arnoldi's method due to the singularity. Therefore, a transformation is needed, which also serves as a preconditioner for the eigenvalue problem. In a first step, a shift-invert transformation with zero shift is applied, which is defined as

$$T_{\text{SI}}(\varsigma_1) := (\boldsymbol{A} - \varsigma_1 \boldsymbol{B}) \cdot \hat{\boldsymbol{x}} = (\gamma - \varsigma_1) \boldsymbol{B} \cdot \hat{\boldsymbol{x}}$$
$$\Rightarrow \theta \hat{\boldsymbol{x}} = (\boldsymbol{A} - \varsigma_1 \boldsymbol{B})^{-1} \cdot \boldsymbol{B} \cdot \hat{\boldsymbol{x}} \quad (3.4.2)$$
$$\Rightarrow \theta \hat{\boldsymbol{x}} = \boldsymbol{A}^{-1} \cdot \boldsymbol{B} \cdot \hat{\boldsymbol{x}} \quad \text{for} \quad \varsigma_1 = 0$$

with $\theta = (\gamma - \varsigma_1)^{-1}$. With the shift-invert transformation, the spurious eigenvalues are mapped to $\Re(\theta) \to 0$ and the eigenvalues close to the shift $\varsigma_1 = 0$ are detected, thus giving the eigenvalues with smallest absolute value (see Bai et al. (2000)). According to Lehoucq & Scott (1997), this step also acts as a purification process, in which undesired null spaces from Arnoldi vectors are removed. The null space is that eigenspace, which corresponds to the infinite eigenvalues.

The shift-invert strategy with zero shift will detect the critical eigenvalue for stationary and oscillatory modes with small absolute values. To ensure that no eigenvalue with smallest real part is missed, a Cayley transformation is performed in a second step, where the eigenvector, corresponding to the eigenvalue with smallest absolute value, of the shift-invert step serves as a starting vector. The Cayley transformation is given by

$$T_{\text{C}}(\varsigma_1, \varsigma_2) := (\boldsymbol{A} - \varsigma_2 \boldsymbol{B}) \cdot \hat{\boldsymbol{x}} = \theta (\boldsymbol{A} - \varsigma_1 \boldsymbol{B}) \cdot \hat{\boldsymbol{x}}$$
$$\Rightarrow (\boldsymbol{A} - \varsigma_1 \boldsymbol{B})^{-1} \cdot (\boldsymbol{A} - \varsigma_2 \boldsymbol{B}) \cdot \hat{\boldsymbol{x}} = \theta \hat{\boldsymbol{x}}. \quad (3.4.3)$$

Now the eigenvalues at infinity are transformed to $\Re(\theta) \to 1$ and the original eigenvalues are obtained by $\gamma = (\varsigma_1 \theta - \varsigma_2)/(\theta - 1)$. According to Lehoucq & Salinger (2001), the Cayley system (3.4.3) has a smaller condition number (ratio of largest-to-smallest eigenvalue in magnitude) than the shift-invert one (3.4.2). Thus, the Cayley transformation results in a better conditioned set of linear equations. The Cayley transformation serves as a verification process for the shift-invert mapping and guarantees that no eigenvalues with smallest real part are missed.

A common approach for choosing the two free Cayley parameters is $\varsigma_1 + \varsigma_2 = 2\gamma_{\text{ref}}$, see Cliffe et al. (1993), Seydel (1994), Meerbergen et al. (1994) and Lehoucq & Scott (1997), which is also adopted here. However, a second equation is needed to determine ς_1 and ς_2, and γ_{ref} has to be specified. Several choices for the two Cayley parameters as suggested in the literature have been tested. The best results with respect to the robustness and convergence rate were obtained by the recommendation of Meerbergen et al. (1994). Assume that we want to computed p

eigenvalues of the system (3.4.3) and we have at least $p+1$ eigenvalues at hand from a previous shift-invert transformation. These eigenvalues are sorted by their real part in ascending order, i.e. $\Re(\gamma_1) < \Re(\gamma_{p+1})$. Then the eigenvalues of (3.4.2) are used to select the Cayley parameters by solving

$$\frac{|\Re(\gamma_1) - \varsigma_2|}{|\Re(\gamma_1) - \varsigma_1|} = \rho_{\text{emp}} \tag{3.4.4a}$$

$$\varsigma_1 + \varsigma_2 = 2\underbrace{\Re(\gamma_{p+1})}_{\gamma_{\text{ref}}}. \tag{3.4.4b}$$

Solving (3.4.4a), for instance, for ς_1 satisfying $\varsigma_1 < \Re(\gamma_1)$, as we are interested in the eigenvalue with smallest real part, the shift is obtained as

$$\varsigma_1 = \frac{\Re(\gamma_1) - 2\Re(\gamma_{p+1}) + \rho_{\text{emp}}\Re(\gamma_1)}{\rho_{\text{emp}} - 1}, \tag{3.4.5}$$

where the empirical parameter is typically set to $\rho_{\text{emp}} \in \{1.2, 1.5\}$ (Meerbergen *et al.*, 1994). For the present type of problems, the value $\rho_{\text{emp}} = 1.5$ has proven to be a good choice. Note that for the Arnoldi method the Krylov-subspace dimension $\mathcal{K}$ has to be at least $\mathcal{K} > 2p$, if one wants to compute p eigenvalues. A higher Krylov-subspace dimension should guarantee that no dangerous, especially critical, eigenvalue is missed in the approximation of the spectrum.

3.5. Algorithms for Root-Finding and Minimization

The optimization problem (2.2.9) evaluates to zero at the critical conditions, i.e.

$$\min_{k \in \mathbb{R}} \sigma\left(k_c, \text{Re}_c, \hat{\boldsymbol{x}}\right) = 0. \tag{3.5.1}$$

Determining the critical parameters constitutes a root-finding problem (Re_c) in combination with a minimization problem (k_c). (3.5.1) can be reformulated to

$$g\left(\text{Re}_c, k_c\right) = 0. \tag{3.5.2}$$

Here it is assumed that the eigenvalue-detection algorithm, as described in the previous section, returns the most dangerous eigenvalue over all discrete eigenvectors $\hat{\boldsymbol{x}} = (\hat{\boldsymbol{u}}, \hat{p})^{\text{T}}$. Therefore, the parameter $\hat{\boldsymbol{x}}$ has been dropped in (3.5.2). The parameter k_c is determined by the minimization problem $k_c = \min_{k \in \mathbb{R}} \sigma(k)$, which gives the most dangerous eigenvalue over all positive wave numbers $k \in \mathbb{R}$. By neglecting the minimization problem, (3.5.2) can be regarded as a single root-finding task $h(\text{Re}_c) = 0$.

Brent's method (Brent (1973)) constitutes a very robust root-finding algorithm, combining the bisection method, the secant method and inverse quadratic interpolation. The bisection method converges for sure to the root, but only linearly. On the other hand, the secant method and the inverse quadratic interpolation converge faster with an order of convergence of about 1.62 and 1.84, respectively. However, there is no guarantee that both of them will converge, if the initial iterates are not very close to the actual root. This does not pose any problem as long as the increment in the Reynolds number Re_{inc} is not too big. Brent's method is very conservative in accepting the value of inverse quadratic interpolation and often calls the bisection method.

In this work, the roots are sought by means of a secant method in conjunction with inverse quadratic interpolation, representing a modification to Brent's method. The procedure is given in the algorithm 3.1, where Re_{beg} represents a start value corresponding to a stable flow condition ($\sigma > 0$) and Re_{inc} is a relative increment in the Reynolds number, such as $\text{Re}_{\text{inc}} = 1.2$.

Algorithm 3.1 Finding the roots of $h(\text{Re}_c) = 0$

$a = \text{Re}_{\text{beg}}; \quad ha = h(a); \quad b = a * \text{Re}_{\text{inc}}; \quad hb = h(b);$
while $ha * hb > 0$ **do**
$\quad a = b; \quad ha = hb; \quad b = a * \text{Re}_{\text{inc}}; \quad hb = h(b);$
end while
{// Root is between a and $b \rightarrow$ secant method}
$c = b - (b-a) * hb/(hb - ha); \quad hc = h(c);$
for $i = 1 \ldots maxiter$ **do**
$\quad$ {// inverse quadratic interpolation}
$\quad d = a * hb * hc/[(ha - hb) * (ha - hc)] + b * ha * hc/[(hb - ha) * (hb - hc)] \ldots$
$\quad \quad + c * hb * ha/[(hc - hb) * (hc - ha)];$
$\quad hd = h(d);$
$\quad$ **if** $|d - c|/c < tol_{\text{Re}}$ **and** $|hd| < tol_\sigma$ **then**
$\quad \quad \text{Re}_c = d; \quad \sigma_c = hd;$
$\quad \quad$ **return** Re_c, σ_c
$\quad$ **end if**
$\quad$ **if** $d < c$ **then**
$\quad \quad b = c; \quad hb = hc; \quad c = d; \quad hc = hd;$
$\quad$ **else**
$\quad \quad a = c; \quad ha = hc; \quad c = d; \quad hc = hd;$
$\quad$ **end if**
end for

The minimization problem $k_c = \min_{k \in \mathbb{R}} \sigma(k)$ is solved by a combination of golden-section search and successive parabolic interpolation. Basic algorithms for the above-mentioned schemes can be found in Brent (1973). The golden-section search is very robust and converges linearly, but slightly faster than the bisection method. The golden-section search divides the interval proportionally to the golden ratio-conjugate $\phi = \varphi^{-1} = \varphi - 1 \approx 0.618$, where $\varphi = (1 + \sqrt{5})/2 \approx 1.618$ defines the golden ratio. Within successive parabolic interpolation (polynomial interpolation of degree two), the minimum of a parabola, which is fitted through three points, is taken. Also its error bounds are calculated in the algorithm 3.2, see Seydel (1994). Successive parabolic interpolation features a superlinear rate of convergence of approximately 1.324.

The procedure for solving the minimization problem is given in the algorithm 3.2, which needs, for a given Reynolds number, a search interval $k \in [a; b]$ as input, in which the minimum lies.

Algorithm 3.2 Minimizing $k_c = \min_{k \in \mathbb{R}} \sigma(k)$ within $k \in [a; b]$

$fa = f(a); \quad fb = f(b); \quad \phi = (1 + \sqrt{5})/2 - 1;$
{// golden section search}
$c = a * \phi + b * (1 - \phi); \quad fc = f(c);$
$d = b * \phi + a * (1 - \phi); \quad fd = f(d);$
if $fc < fd$ **then**
  $b = d; \quad fb = fd;$
else
  $a = c; \quad fa = fc; \quad c = d; \quad fc = fd;$
end if
{// actual minimum is saved in the variable c ($fc < fa \wedge fc < fb$)}
for $i = 1 \ldots maxiter$ **do**
  {// successive parabolic interpolation d_0 with error bounds $d_{1,2}$}
  $d_0 = b - 0.5 * \{[(b-a)^2 * (fb - fc) - (b-c)^2 * (fb - fa)] \ldots$
       $/[(b-a) * (fb - fc) - (b-c) * (fb - fa)]\};$
  $\xi_1 = (fb - fc)/(b - c); \quad \xi_2 = [(fa - fc)/(a - c) - \xi_1]/(a - b);$
  $d_{1,2} = c + \left\{-\xi_1 - \xi_2 * (c - b) \pm \sqrt{[\xi_1 + \xi_2 * (c - b)]^2 - 4 * \xi_2 * fc}\right\}/(2 * \xi_2);$
  $d \in \{d_0, d_1, d_2\};$
  if $d \notin [(3a + c)/4; (3b + c)/4]$ **then**
    {// golden section search}
    if $fa < fb$ **then**
      $d = c * \phi + a * (1 - \phi);$
    else
      $d = c * \phi + b * (1 - \phi);$
    end if
  end if
  $fd = f(d);$
  if $|d - c|/c < tol_k$ **then**
    $k = d;$
    return k
  end if
  if $d < c$ **then**
    $b = c; \quad fb = fc; \quad c = d; \quad fc = fd;$
  else
    $a = c; \quad fa = fc; \quad c = d; \quad fc = fd;$
  end if
end for

3.6. Plane Poiseuille Flow

As a test case of the numerical implementations, the two-dimensional plane channel flow is considered with periodic boundary conditions in the streamwise (x) direction $\boldsymbol{u}_0(x = 0, y) = \boldsymbol{u}_0(x = L, y)$. The flow is driven by a pressure gradient over the channel length L

$$\frac{\mathrm{d}p}{\mathrm{d}L} = -\frac{4U_\infty^2}{H\mathrm{Re}} = -\frac{2}{\mathrm{Re}}, \qquad (3.6.1)$$

where the Reynolds number Re is based on the centreline velocity ($U_\infty = 1$) and half the channel height ($H = 2$). The numerical solution converges to the analytical plane Poiseuille profile

$$u_0(y) = \frac{4U_\infty}{H^2} y(H - y). \qquad (3.6.2)$$

Considering two-dimensional perturbations

$$\begin{pmatrix} \tilde{u} \\ \tilde{v} \\ \tilde{p} \end{pmatrix}(x, y, t) = \begin{pmatrix} \hat{u} \\ \hat{v} \\ \hat{p} \end{pmatrix}(x, y)\, \mathrm{e}^{-\gamma t} + \mathrm{c.c.} \qquad (3.6.3)$$

with periodic boundary conditions (also for $\tilde{p}$) in the streamwise direction, Tollmien–Schlichting waves were obtained at the same critical parameters as in Thomas (1953) and Orszag (1971), see table 3.1. These two authors did not solve the two-dimensional Navier–Stokes, but the Orr–Sommerfeld equation and applied an exponential ansatz for the x direction.

authors	Re_c	k_c	$|\omega_c|$
Thomas (1953)	5780	1.026	0.265
Orszag (1971)	5772	1.021	0.264
present	5779	1.021	0.269

Table 3.1.: Comparison of critical Reynolds numbers Re_c, wave numbers k_c and oscillatory frequencies ω_c with previous numerical results.

Figure 3.2 shows the streamlines of the perturbations and the total local energy transfer at critical conditions, i.e. $\mathrm{Re}_c = 5779$ and $L_c = 2\pi/1.021$ with $\omega_c = 0.269$. From figure 3.3 it can be noticed that $\tilde{u}$ and $\tilde{p}$ are antisymmetric (odd), whereas $\tilde{v}$ is symmetric (even) with respect to $x = \pi/k_c$.

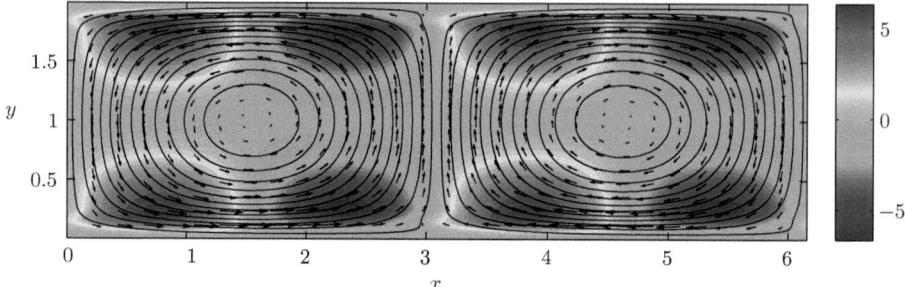

Figure 3.2.: Streamlines of the critical mode, perturbation flow (arrows) and the total local energy production $\sum_i I_i = I'_2$ at critical conditions. The Tollmien–Schlichting wave travels to the left.

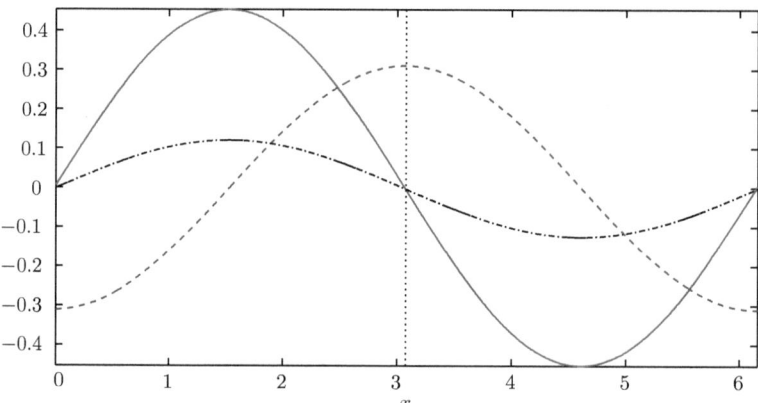

Figure 3.3.: The perturbation components, integrated up to half the channel height $H/2 = 1$, as functions of the x direction. The red solid, the blue dashed and the black dash-dotted line represent $\int_0^1 \tilde{u}\,dy$, $\int_0^1 \tilde{v}\,dy$ and $\int_0^1 \tilde{p}\,dy$, respectively. The black dotted line denotes $x = \pi/k_c$.

4. Results

It is worth mentioning that the results of the sections 4.1, 4.2 and 4.3 have already been published in the Journal of Fluid Mechanics. The papers in question are Lanzerstorfer & Kuhlmann (2012a), Lanzerstorfer & Kuhlmann (2012b) and Lanzerstorfer & Kuhlmann (2012c), respectively.

4.1. The Backward-Facing-Step Problem

4.1.1. Problem Formulation

The incompressible flow of a Newtonian fluid over a backward-facing step in the (x, y) plane is considered. The geometry is sketched in figure 4.1. It consists of an inlet channel of height h_i and length L_i, followed by a suddenly enlarged channel of height H and length L_o. The origin of the Cartesian coordinate system is located at the bottom of the step, which has a height $h_s = H - h_i$. The geometry is characterized by the non-dimensional expansion ratio $\Gamma^b = h_s/H$. The system is assumed to be homogeneous and infinitely extended in the spanwise (z) direction. This assumption simplifies the numerical analysis and allows one to study intrinsic effects, unmasked by sidewall perturbations.

The Reynolds number

$$\mathrm{Re} = \frac{LU_\infty}{\nu} \qquad (4.1.1)$$

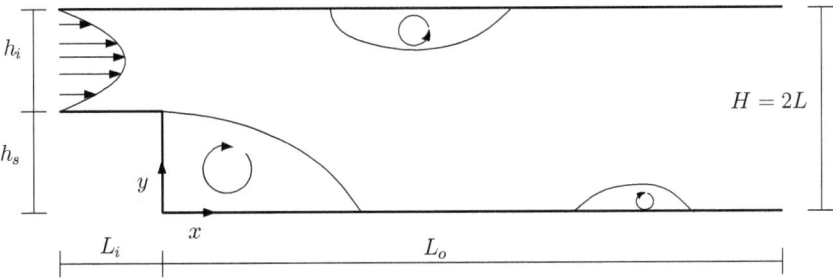

Figure 4.1.: Sketch of the flow geometry for the backward-facing-step problem.

is based on half of the outlet channel height $L = H/2$ and the centreline (maximum) velocity U_∞ of the plane Poiseuille flow, which is imposed at the inflow boundary. Length, velocity and pressure of the steady Navier–Stokes equations (2.1.5) have been made dimensionless with respect to L, U_∞ and ρU_∞^2.

At the inlet boundary, plane Poiseuille flow is prescribed as $u_0 = 4(y - h_s)(H - y)/h_i^2$ and $v_0 = 0$ for $y \in [h_s; H]$. Along the step and all channel walls no-slip and no-penetration boundary conditions $\boldsymbol{u}_0 = 0$ are imposed. At the outlet, the standard outflow boundary condition $\partial_x \boldsymbol{u}_0 = 0$ and $p_0 = 0$ is employed (see e.g. Gartling, 1990). These outflow boundary conditions are of absorbing and non-reflecting type in the case of stationary simulations (Ol'shanskii & Staroverov, 2000).

4.1.2. Scientific Background

The two-dimensional flow over a backward-facing step has been subject to several numerical investigations (e.g. Kim & Moin (1985), Sohn (1988), Gartling (1990), Gresho et al. (1993)) and often serves as a benchmark problem for numerical codes (see e.g. Blackwell & Pepper (1992)). However, like in many other studies, a plane Poiseuille flow is prescribed right at the step without including an inlet channel. Such a computational domain does not have any step corner and the solution deviates from the one observed in the experiments. Without an upstream channel, the sharp corner singularities do not arise, which may otherwise have a significant effect on the flow in the vicinity of the step. Moffatt (1964) developed a local asymptotic expansion for the flow around a right-angled corner, but his creeping-flow solution is only valid in the immediate vicinity of the corner. Hawa & Rusak (2002) showed that several levels of local mesh refinement are required to accurately reproduce the asymptotic solution of Moffatt (1964) by numerical simulation in the vicinity of the corner.

Cruchaga (1998) solved the two-dimensional backward-facing-step problem with and without using an inlet channel and found that the two solutions differ substantially in the vertical velocity component and the pressure. Moreover, Erturk (2008) compared the two-dimensional basic flow (u_0, v_0) obtained for a long inlet channel with the benchmark solution of Gartling (1990) that did not include any upstream channel. As a result, it was found that the flow quantities v_0, $\partial_x v_0$ and $\partial_x u_0$ are very sensitive to the presence/absence of an upstream channel, in particular, in the vicinity of the step.

Based on numerical simulations for a spanwise-periodic domain with an expansion ratio $\Gamma^b = 0.5$, i.e. the ratio of the step height to the channel height downstream, Kaiktsis et al. (1991) reported both two- and three-dimensional instabilities at $\text{Re} \approx 525$ in contradiction to the experimental data of Armaly et al. (1983) and the numerical results of Gartling (1990). This discrepancy caused a great controversy and led Gresho et al. (1993) to publish a study applying four different methods: time-marching finite-element, finite-difference and spectral-element methods and a steady finite-element method followed by a two-dimensional linear stability analysis. They concluded that the steady flow is stable at Re = 600 and that the results of Kaiktsis et al. (1991) were false owing to inaccuracy of their numerical scheme. Kaiktsis et al. (1996) admitted the incorrectness of their former simulations and showed that the steady flow is linearly stable against two-dimensional perturbations at least up to Re = 1875, the maximum value covered. Moreover, they argued that the flow is convectively unstable for Re $\geq$ 525. Fortin et al. (1997) confirmed that the basic two-dimensional flow is linearly stable with respect to two-dimensional perturbations up to Re = 1200, the maximum Reynolds number covered.

More recently, Yanase et al. (2001) performed direct numerical simulations for a geometry without an inlet channel and for an expansion ratio of $\Gamma^b = 0.5$, assuming periodic boundary conditions in the spanwise direction. Computations were carried out for moderate Reynolds numbers Re $\leq$ 700. For 525 $\leq$ Re $\leq$ 700, transient growth of the perturbations was detected, which, however, decayed for $t > 70$ convective time units. They argued that the three-dimensional vortical structures, triggered by imposing small inlet disturbances, are directly related to the high shear stresses between the bulk flow and the primary vortex behind the step, which might be traced back to the Kelvin–Helmholtz instability. Convective instabilities were also studied by Blackburn et al. (2008) for an expansion ratio of $\Gamma^b = 0.5$ by a transient-growth analysis. They stated that for Re < 57.7 there exists no energy growth over any time interval. Besides that, optimal three-dimensional disturbances, resulting in the highest growth rate, were identified with a spanwise wavelength of the order of 10 step heights.

Barkley et al. (2002) performed a three-dimensional linear stability analysis of the two-dimensional flow for a geometry with an expansion ratio $\Gamma^b = 0.5$ and an entrance length of one step height. They claimed that the base flow is linearly stable up to a critical Reynolds number of 748 and that the centrifugal-instability mechanism is responsible for generating the three-dimensional flow.

The convective and two-dimensional, temporal linear stability of a flow over a backward-facing step has been studied extensively in the literature. To the best of the authors' knowledge, Barkley et al. (2002) were the only authors to perform a global (three-dimensional) stability analysis, but merely for the single expansion ratio of $\Gamma^b = 0.5$. In the present work also a global linear stability analysis of the two-dimensional basic flow is carried out with the objective to systematically vary the geometry, where the expansion ratio is covered from $\Gamma^b = 0.25$ to $\Gamma^b = 0.975$. It will be shown that the stability boundary of Barkley et al. (2002) does not apply to the ideal case of an infinitely long entrance channel, since the entrance channel used was too short. Since a spanwise-periodic computational domain is addressed, extrinsic, three-dimensional effects induced by no-slip sidewalls are excluded and the fundamental hydrodynamic instability mechanism can be studied. As it will turn out, the physical nature of the instability is neither a centrifugal, as argued by Barkley et al. (2002), nor a Taylor–Görtler instability mechanism as proposed by Ghia et al. (1989). The instability is rather a combina-

L_i	1L	2L	4L	5L	10L	20L
Re_c	748.3	717.8	713.7	713.9	714.05	714.1
k_c	0.9155	0.8852	0.8754	0.8758	0.8778	0.8767

Table 4.1.: Critical values as functions of the entrance length for multiples of the length scale L and for $\Gamma^b = 0.5$ with $L_o = 60L$.

tion of flow deceleration, lift-up process and streamline convergence for an expansion ratio of $\Gamma^b = 0.5$. For other geometries the lift-up mechanism, pure elliptical and centrifugal amplification processes will be identified. All these elementary processes will be explained later in the text.

4.1.3. Results

Parameter dependence

To verify the results of the three-dimensional linear stability and energy analyses, the code is employed to compute the critical Reynolds number and its associated energy-transfer rates for the lid-driven square cavity. The results obtained, $\text{Re}_c = 786.1$ (786.3), $k_c = 15.42$ (15.43), $\int_V I'_1 dV = 0.04$ (0.04), $\int_V I'_2 dV = 0.68$ (0.68), $\int_V I'_3 dV = 0.18$ (0.18) and $\int_V I'_4 dV = 0.1$ (0.1) agree with the data of Albensoeder et al. (2001) given in parentheses. Also the calculations of Theofilis (2000) ($\text{Re}_c = 783$, $k_c = 15.4$) are consistent with our results.

Another check is provided by a comparison of the critical Reynolds and wave number with the results of Barkley et al. (2002) for the backward-facing-step problem, where three-dimensional perturbations were considered. Using a spectral-element code they obtained, for $\Gamma^b = 0.5$ and an inlet channel of length $1L$, $\text{Re}_c = 748$ and $k_c = 0.91$. These data fit well to the present results given in table 4.1. From table 4.1 it is also clearly evident that the entrance length used by Barkley et al. (2002) is too short if entrance-length-independent results are desired. Note that Barkley et al. (2002) utilized an outflow-channel length of $L_o^{\text{Bar}} = 35L$, which is long enough for not influencing the stability boundaries.

To complete the verification of the discretization of the bulk equations, extensive grid-convergence studies have been carried out. The highest possible resolution was 2800×200 finite volumes in the (x, y) plane, corresponding to approximately 2.24 million unknowns. On this uniform grid with a spacing of $\Delta x = 0.025$ and $\Delta y = 0.01$, the benchmark solution $\text{Re}_c = 714.55$ and $k_c = 0.8787$ is obtained for $\Gamma^b = 0.5$ with $L_i = 10L$ and $L_o = 60L$. On a geometrically refined grid only about one-third of the cells is required. The same results as the benchmark solution with a deviation of less than 0.2 % are obtained with three different grids and their parameters are summarized in table 4.2. Here N_x^i and N_y^i represent the number of cells in the inlet channel in the x and y directions and N_x^o and N_y^o those in the outlet channel, respectively.

	$\Delta x_{\min}$	$\Delta y_{\min}$	$\Delta x_{\max}$	$\Delta y_{\max}$	N_x^i	N_y^i	N_x^o	N_y^o
G_1	0.015	0.005	0.06	0.02	188	94	1019	188
G_2	0.0112	0.0037	0.09	0.03	153	110	714	220
G_3	0.0187	0.0063	0.075	0.025	155	84	820	168

Table 4.2.: Three different grids with constant stretching factor of 1.03 for $\Gamma^b = 0.5$ with $L_o = 60L$.

A detail around the corner of grid G_3 of table 4.2 is depicted in figure 4.2 for $\Gamma^b = 0.5$. As this grid (G_3) is the most efficient one, it is used in almost all the calculations, except for

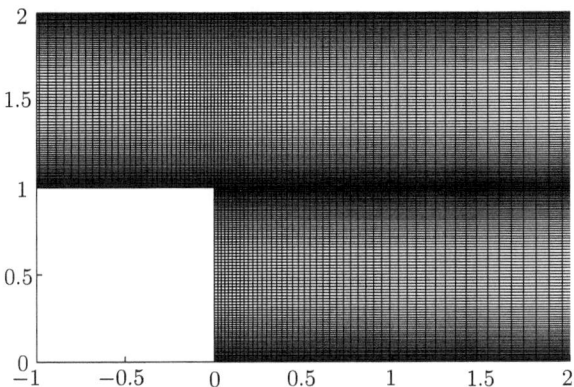

Figure 4.2.: Sketch of the grid G_3.

$\Gamma^b = 0.975$, where the grid G_2 is utilized as it features more control volumes in the y direction. The grid convergence presented here for $\Gamma^b = 0.5$ similarly holds for all the other expansion ratios covered in this work.

For the generation of the various grids for different expansion ratios Γ^b, the minimum grid spacing at the walls $\Delta x_{\min}$ and $\Delta y_{\min}$ was kept fixed. The cell size is then increased with a constant stretching factor of 1.03 until a maximum value $\Delta x_{\max}$ or $\Delta y_{\max}$, respectively, is obtained, or the end of the computational domain is reached. If the maximum cell size is reached it is kept constant up to the boundaries of the geometry. The grid parameters are given in table 4.3, apart from N_x^i, which can be found in table 4.2.

The entrance length L_i and the length of the channel L_o should be large enough such that the results are independent of these two parameters. The required outflow length L_o depends on the Reynolds number and it has to be sufficiently long so that a plane Poiseuille flow is achieved at the outlet. Here the outflow length L_o is chosen in such a way that the maximum relative deviation of the numerical solution from the fully developed plane Poiseuille flow is

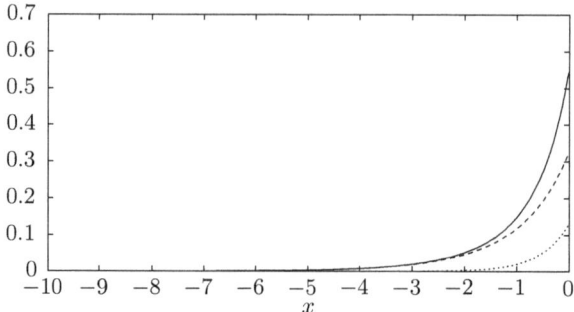

Figure 4.3.: The relative amplitudes of the perturbations in the inlet channel averaged over the spanwise direction $\text{mean}_z \max_y |\tilde{\boldsymbol{u}}(x,y,z)|$ for $\Gamma^b = 0.5$. The solid, the dotted and the dashed line represent the $\tilde{u}$, the $\tilde{v}$ and the $\tilde{w}$ component, respectively.

less than 3%. With this choice $L_o = 60L$ for $\Gamma^b = 0.5$, which is almost twice the length of Barkley *et al.* (2002)'s geometry with $L_o^{\text{Bar}} = 35L$.

For $\Gamma^b = 0.5$, Mateescu & Venditti (2001) concluded that an inlet channel of length $4L$ is sufficient to obtain entrance-length-independent basic flows. As this thesis is concerned with a linear stability analysis, it is important that the critical Reynolds Re_c and wave numbers k_c are independent of the entrance length. The dependence of these critical values on the length of the inlet channel L_i is provided in table 4.1. The results are given for $\Gamma^b = 0.5$, but a similar convergence, as L_i is increased, holds for all expansion ratios covered here. An inlet channel with length $L_i = 10L$ is sufficient, given that the algorithm by itself has a tolerance level of about 0.2% for each critical data. A sufficiently long inlet channel is required because the critical modes extend upstream of the step. The relative amplitudes of the perturbation components in the inlet channel averaged over the spanwise direction $\text{mean}_z \max_y |\tilde{\boldsymbol{u}}(x,y,z)|$ are depicted in figure 4.3 for $\Gamma^b = 0.5$. Note that the amplitudes are normalized to sum up to one.

To sum up, all the results presented in this work are computed with an entrance length $L_i = 10L$ and a channel length $L_o \in [35; 400]$ depending on the Reynolds number (see table 4.3). All the calculations are performed on the grid G_3, except for $\Gamma^b = 0.975$, where the grid G_2 is used. The maximum stretching factor is not larger than 1.03 which ensures a nearly second-order discretization.

Stability boundaries

Representative numerical critical Reynolds and wave numbers are collected in table 4.3 for selected expansion ratios. Also the outflow lengths and the number of grid points used are indicated.

Γ^b	Re_c	k_c	$\pm\omega_c$	L_o	N_y^i	N_x^o	N_y^o
0.25	4912.8	1.2477	0	400	104	5339	158
0.3	2948.0	1.0156	0	300	100	4010	162
0.4	1288.1	0.8611	0	90	92	1218	166
0.5	714.05	0.8778	0	60	84	820	168
0.6	488.05	0.9952	0	40	74	554	166
0.7	402.37	1.3062	0	35	62	487	162
0.8	362.67	1.3883	0.0279	35	46	487	154
0.9	404.67	1.4770	0.0483	35	28	487	144
0.95	543.73	1.6675	0.0506	40	16	554	136
0.975	819.08	1.8446	0.0441	40	14	490	162

Table 4.3.: Critical parameters for selected expansion ratios Γ^b, the outflow length L_o and the number of grid points (N_x, N_y) in the (x, y) plane.

Critical Reynolds and wave numbers are shown in figure 4.4 as functions of the expansion ratio Γ^b for $\Gamma^b \leq 0.5$. In the range shown, the instability is stationary and the critical curves vary smoothly with Γ^b.

When the step height is reduced to small expansion ratios, the critical Reynolds number increases monotonically. For the smallest step height considered here, i.e. for $\Gamma^b = 0.25$, the critical wave and Reynolds numbers are $\mathrm{Re}_c = 4912.8$ and $k_c = 1.2477$, respectively. This corresponds to a Reynolds number in the inlet channel (maximum velocity and half the height of the inlet channel) of $\mathrm{Re}_{\mathrm{inlet}} = 3684.6$, which is below the linear stability boundary of plane Poiseuille flow.

Additional calculations with a Krylov-subspace dimension of 500 were performed to ensure that no eigenmodes are missed. For all expansion ratios covered, a Krylov-subspace dimension of 500 yields the same results as a Krylov-subspace dimension of 200. An example is given in figure 4.5 for $\Gamma^b = 0.25$.

The critical curves for higher expansion ratios $\Gamma^b \geq 0.5$ are shown in figure 4.6. At $\Gamma^b_{co} = 0.7090$ the critical mode changes from stationary (for $\Gamma \leq \Gamma^b_{co}$) to oscillatory (for $\Gamma^b > \Gamma^b_{co}$). At the crossover point Γ^b_{co} two real eigenvalues are annihilated and a pair of complex conjugate ones emerges. The critical wave number does not change very much. The merger of the complex conjugate eigenvalues occurs along a line on the two-dimensional neutral surface in the (k, Γ^b) space. At Γ^b_{co} this line is close to the minimum with respect to k of the neutral hypersurface. The behaviour is illustrated in figure 4.7 where the two most dangerous eigenvalues are shown along a ray $(\Gamma, \mathrm{Re}) = (0.7090, 401.09)$ in the three-dimensional parameter space spanned by Γ, Re and k.

The critical Reynolds number of the oscillatory mode takes a minimum value at $\Gamma^b = 0.8291$ with $\mathrm{Re}_c = 359.06$ and $k_c = 1.3919$. In the limit $\Gamma \to 1$ the step height approaches the width

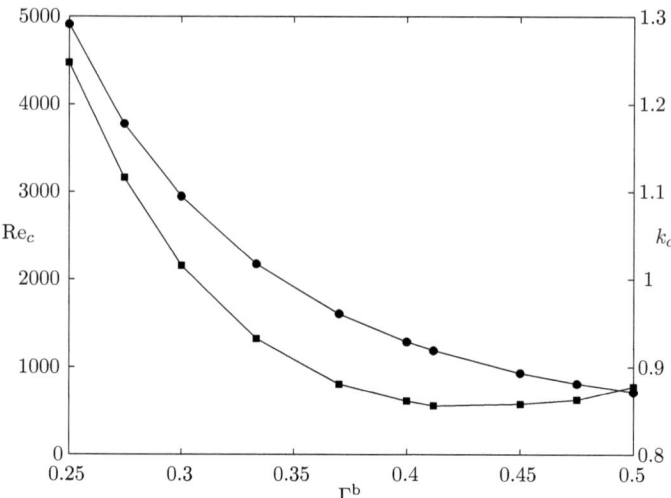

Figure 4.4.: Critical Reynolds number Re_c (dots) and wave number k_c (squares) as functions of the expansion ratio Γ^b.

of the outlet channel. In this limit the basic flow becomes similar to a wall jet and the length scale $L = (h_i + h_s)/2$ may not be the best choice. If the width of the inlet channel h_i is used as a length scale, the Reynolds number, wave number and frequency must be rescaled according to

$$Re^* = 2(1-\Gamma)Re, \quad k^* = 2(1-\Gamma)k, \quad \text{and} \quad \omega^* = 2(1-\Gamma)\omega. \tag{4.1.2}$$

For the largest expansion ratio $\Gamma^b = 0.975$ considered, the critical data $Re_c^* = 40.954$ with $k_c^* = 0.0922$ and $\omega_c^* = 0.0022$ are obtained. As can be seen from figure 4.8, Re_c^* shows a linear behaviour, which is estimated, for $(1-\Gamma) \ll 1$, as $Re_c^* \approx 27 - 500(\Gamma^b - 1)$. The critical wave number k_c^* for $\Gamma^b \to 1$ becomes very small and can be approximated by $k_c^* \approx 2.9051 - 2.8842\Gamma$. The frequencies ω_c and ω_c^* of the critical mode for large expansion ratio ($\Gamma^b \geq 0.7$) are depicted in figure 4.9 with an estimate of the asymptote of $\omega_c^* \approx -0.0882(\Gamma - 1)$.

The stationary and the oscillatory critical modes are continuous functions of the expansion ratio Γ^b. Even at the crossover point Γ_{co}^b the two critical Reynolds numbers are very similar because they are closely connected to each other in the parameter space. Nevertheless, the structure of the critical mode and, hence, the instability mechanism changes. For that reason, representative expansion ratios are considered and the critical modes are analysed towards a physical interpretation of the instability mechanisms. In this regard the kinetic energy exchanged between the basic state and the perturbation field can provide useful information. For an overview, the normalized energy production terms integrated over the whole domain of the flow are shown in table 4.4 for selected expansion ratios. Note that the sum of all normalized

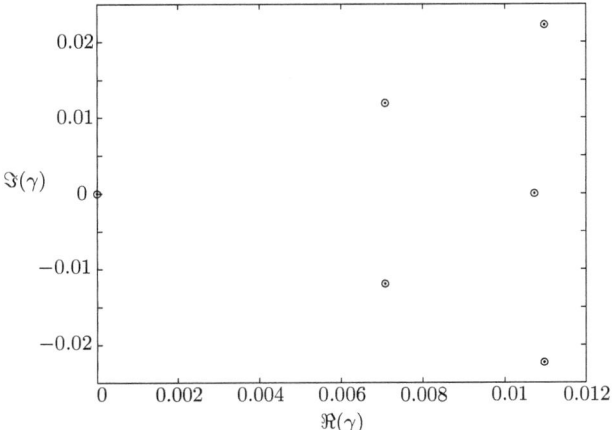

Figure 4.5.: Part of the eigenspectrum showing the six most dangerous eigenvalues for a Krylov-subspace dimension of 200 (open circles) and 500 (dots) for $\Gamma^b = 0.25$ at critical conditions.

energy production terms equals one up to 4 to 5 significant digits. The last row $\int_{S_o}$ in table 4.4 constitutes the surface integral at the outlet $-0.5 \int_{S_o} I_5 \mathrm{d}S$.

For $\Gamma^b = 0.25$ the perturbations are not fully decayed at the outlet owing to the finite length of the channel. Performing the same calculation on a coarser grid with a much longer outlet channel indicates that the surface integral vanishes for $L_o \to \infty$.

The instability mechanisms discussed further below will be denoted centrifugal, elliptic and lift-up instability, which have been introduced for simple, elementary basic flows. A rigorous classification of the instabilities is difficult, because no exact criteria exist for general viscous flows in complex geometries. Moreover, the instability characteristics may change continuously upon variation of the expansion ratio. Thus several of the elementary instability mechanisms may be at work simultaneously. Therefore, the instability mechanism will be characterized here by a qualitative and quantitative comparison of the basic flow, the critical mode and its energy-transfer rates with the instabilities arising in elementary idealized systems.

Regardless of the step height the basic flow behind the backward-facing step is characterized by a region of separated flow in the form of a more or less strained vortex immediately behind the step. This vortex will be called the primary vortex. Depending on the Reynolds number, secondary and higher-order regions of separation arise further downstream, alternatingly located on both walls of the channel. The energy transfer between the basic flow and the critical modes is most pronounced near the primary vortex. Hence, the characteristics of the primary separation region is of key importance for the flow instability. Therefore, the flow characteristics shall be discussed within the region of the primary vortex, even though the computational domain is much more extended in the downstream (x) direction.

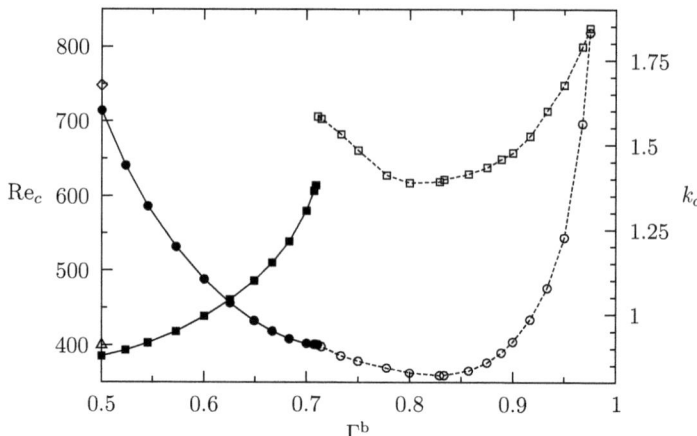

Figure 4.6.: Critical Reynolds number Re_c (dots) and wave number k_c (squares) as functions of the expansion ratio Γ^b. Data for the stationary instability are indicated by full symbols and full lines; those for the oscillatory instability are shown as open symbols and dashed lines. The open diamond and open triangle represent the critical Reynolds number and wave number, respectively, of Barkley et al. (2002).

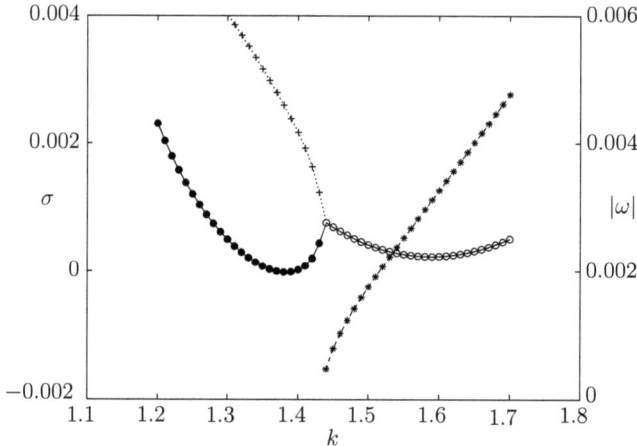

Figure 4.7.: Real and imaginary parts of the two most dangerous eigenvalues as functions of k for $\Gamma^b = 0.7090 \approx \Gamma^b_{co}$ and $\mathrm{Re} = 401.09 \approx \mathrm{Re}_c$. The real parts of the leading eigenvalues for the stationary and the oscillatory modes are represented by dots and circles, respectively. Plus signs indicate the real part of the second eigenvalue, which merges with the first one. The absolute values of the imaginary parts of the oscillatory modes are depicted as asterisks.

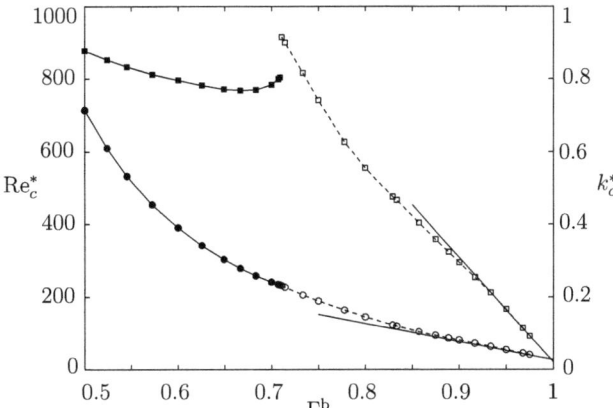

Figure 4.8.: Scaled critical Reynolds numbers Re_c^* (circles) and scaled critical wave numbers k_c^* (squares) as functions of the expansion ratio Γ^b. Data for the stationary instability are indicated by full symbols connected by full lines; those for the oscillatory instability are shown as open symbols and dashed lines. The straight lines indicate the asymptotic behaviour for $\Gamma^b \to 1$ as mentioned in the text.

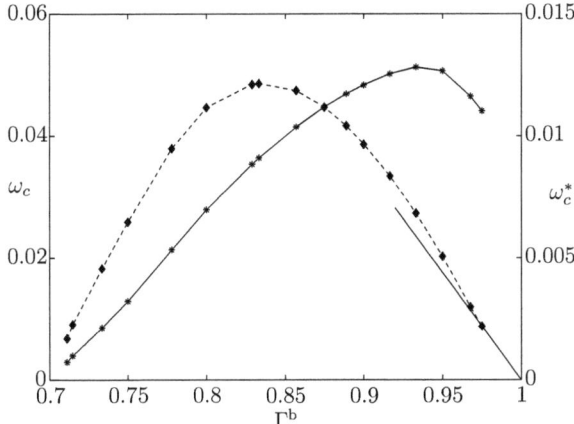

Figure 4.9.: Critical oscillation frequencies ω_c based on U_∞/L (stars, full lines) and ω_c^* based on U_∞/h_i (diamonds, dashed lines) as functions of the expansion ratio Γ^b.

	$\Gamma^b = 0.25$		$\Gamma^b = 0.5$		$\Gamma^b = 0.709$	
i	$\int_V I_i dV$	$\int_V I'_i dV$	$\int_V I_i dV$	$\int_V I'_i dV$	$\int_V I_i dV$	$\int_V I'_i dV$
1	−2.7154	0.3237	−2.5177	0.2670	−1.0081	0.3230
2	3.7858	0.4423	3.5612	0.2992	1.9968	0.5940
3	−0.0045	0.0798	−0.0543	0.0768	−0.0971	0.0146
4	0.0024	0.2226	0.0249	0.3711	0.1114	0.0654
$\int_{S_o}$	−0.0682		−0.0141		−0.0031	

	$\Gamma^b = 0.8$		$\Gamma^b = 0.975$	
i	$\int_V I_i dV$	$\int_V I'_i dV$	$\int_V I_i dV$	$\int_V I'_i dV$
1	−0.2886	0.1946	0.0729	0.2024
2	1.2514	0.5933	0.6133	0.4915
3	−0.0637	0.0485	0.1311	0.1407
4	0.1039	0.1639	0.1828	0.1654
$\int_{S_o}$	−0.0001		0.0000	

Table 4.4.: Global normalized energy production rates for several expansion ratios.

Oscillatory instability for large expansion ratios $\Gamma^b > \Gamma^b_{co}$

Centrifugal instability for very large expansion ratios When the expansion ratio is large a thin plane jet attached to the upper wall emerges from the opening of the inlet channel at $x = 0$. Immediately behind this location the jet resembles a classical wall jet (Batchelor, 1967). It widens downstream and, after a short distance, it separates from the upper wall and is deflected downwards to the lower wall. After being deflected by the lower wall the jet continues to widen. The S-shaped jet is located between the primary and secondary separation eddies. For very large expansion ratios there appears another recirculation bubble which resides at the bottom of the channel.

As a representative example for a large expansion ratio $\Gamma^b = 0.975$ is considered. The primary vortex has nearly circular streamlines in its centre and it is only slightly strained. The basic state and the oscillatory critical mode are shown in figure 4.10 in a cross-section $z = \text{const.}$ in which the total local energy-transfer rate takes its maximum. The critical mode in that plane arises in the form of a vortex, which appears slightly displaced downstream from the primary vortex and centred near the separating streamline. Hence, a finite-amplitude disturbance would be a spanwise travelling wave that periodically displaces the primary eddy and the jet. The total local energy transfer from the basic state to the perturbation flow is sharply peaked at $(x, y) = (3.1230, 1.1141)$ on the downstream side of the jet slightly above the stagnation point on the lower wall.

It is worth mentioning that Chun & Schwarz (1967) examined the temporal linear stability of an incompressible wall jet subject to small disturbances by solving the Orr–Sommerfeld equation. The critical Reynolds number, based on the local boundary layer thickness and the

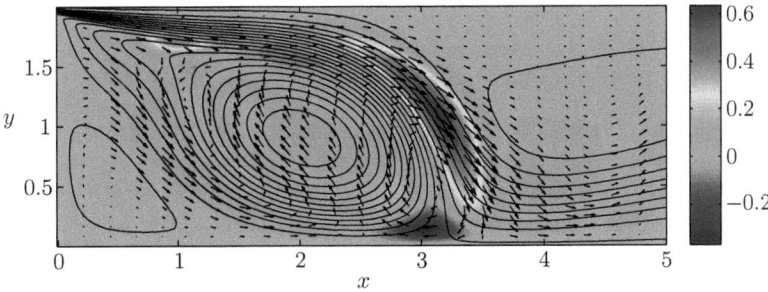

Figure 4.10.: Basic state (streamlines), critical velocity fields (arrows) and the total local energy production $\sum_i I_i$ for $\Gamma^b = 0.975$ at $z = \text{const.}$.

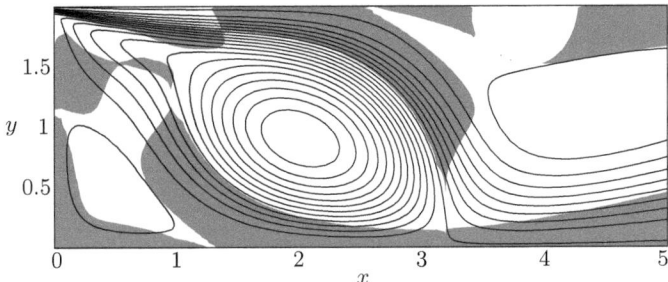

Figure 4.11.: Basic state (streamlines) and grey regions where (4.1.3) holds for $\Gamma^b = 0.975$

local maximum velocity, for the wall jet is $\text{Re}_c^{\text{WJ}} \approx 57$. This value is of the same order of magnitude as the present extrapolated critical Reynolds number $\text{Re}_c^*(\Gamma \to 1) \approx 27$. As the local boundary layer thickness depends on the distance from the inlet channel, the x-positions were determined, where $\text{Re}^{\text{WJ}} \approx 57$ for $\Gamma^b = 0.975$. This condition is fulfilled at $x = 1.1337$ and $x = 3.0854$. These are approximately the positions where the total local energy production takes its maxima (figure 4.10).

In the region of maximum energy transfer the basic streamlines are curved and the separation of the equidistantly plotted streamlines increases radially outwards from the curved streamlines. Hence, this region is prone to centrifugal effects. This is confirmed by figure 4.11, which shows the generalized Rayleigh criterion

$$\frac{|\boldsymbol{u}_0|\zeta}{R} < 0 \qquad (4.1.3)$$

as grey regions. Here ζ represents the vorticity of the steady two-dimensional basic flow $\boldsymbol{u}_0 = (u_0, v_0, 0)^{\mathrm{T}}$ and R the local algebraic radius of curvature which can be expressed as

$$R = \frac{|\boldsymbol{u}_0|^3}{(\nabla \psi) \cdot (\boldsymbol{u}_0 \cdot \nabla \boldsymbol{u}_0)}. \tag{4.1.4}$$

The criterion is a reformulation of Bayly (1988)'s sufficient condition for centrifugal instability for two-dimensional inviscid flows due to Sipp & Jacquin (2000). If (4.1.3) is satisfied all along a closed streamline $\psi = \text{const.}$, an inviscid basic flow $\boldsymbol{u}_0$ would be centrifugally unstable. Here, the criterion is only used as an indicator for centrifugal effects, because the basic flow is viscous and (4.1.3) is only satisfied along certain segments of the streamlines.

Since the most prominent energy-transfer rate I_2' (table 4.4) is compatible with this interpretation, the instability is centrifugal in nature. In the region of large energy transfer the cross-stream component $\tilde{\boldsymbol{u}}_\perp$ is relatively weak compared to $\tilde{\boldsymbol{u}}_\parallel$ (figure 4.10). Nevertheless, a high energy-transfer rate I_2' is achieved due to the high shear rate present at the edge of the curved jet.

Owing to the periodicity in z the regions of positive (acceleration) and negative (deceleration) perturbations $\tilde{\boldsymbol{u}}_\parallel$ is connected by a return flow in the z direction, which is visible in the plane $x = 3.1230$ shown in figure 4.12. Owing to the relatively low critical oscillation frequency ω_c, the obliqueness of the perturbation-flow pattern is weak in figure 4.12.

It is interesting to mention that a similar centrifugal instability occurs in deep lid-driven cavities for cavity aspect ratios $\Gamma^{\mathrm{cav}} > 1.207$ (Albensoeder et al., 2001). In deep cavities the wall jet emerging from the downstream corner of the moving wall also separates from the solid wall and the energy-transfer rate I_2' is peaked in the same region of the jet as in the present case (see e.g. figure 20 of Albensoeder et al., 2001). Moreover, both modes have a relatively

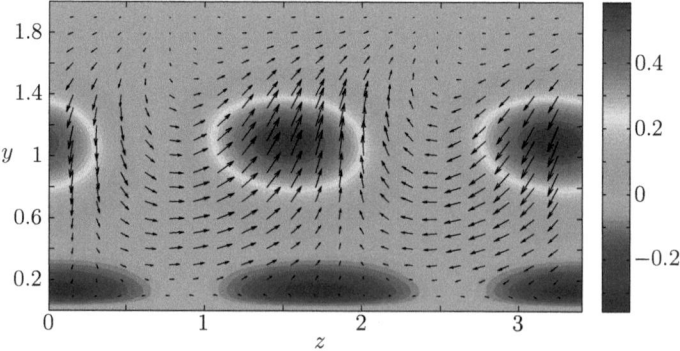

Figure 4.12.: Critical velocity fields (arrows) and the total local energy production $\sum_i I_i$ for $\Gamma^{\mathrm{b}} = 0.975$ at $x = 3.1230$. The wave propagates to the left.

long wavelength, $k_c = 1.845$ as compared to the cavity mode $k_c^{\text{cav}} = 1.715$ for $\Gamma^{\text{cav}} = 2$. The two critical modes differ, however, slightly in shape and in their time dependence: the cavity mode is stationary.

Elliptical instability for moderately large expansion ratios As the expansion ratio is reduced the critical mode changes gradually and the centrifugal instability mechanism is replaced by an elliptic instability mechanism. The elliptic mechanism becomes dominant at about $\Gamma^b \approx 0.9$ and it remains the most dominant mechanism even beyond the critical expansion ratio $\Gamma_{\text{co}}^b < 0.7108$.

The elliptic instability arises in strained vortices (Bayly, 1986; Pierrehumbert, 1986). It is due to an amplification of a pair of Kelvin waves that resonate due to a coupling provided by the strain field (Eloy & Le Dizès, 2001; Kerswell, 2002). The hallmark of the elliptic instability is a peak of the energy-transfer rate in the centre of the basic elliptic vortex (Sipp & Jacquin, 1998) and a perturbation flow in the vortex centre which is aligned with the principal direction of the strain (Waleffe, 1990).

As an example, $\Gamma^b = 0.8$ is considered. At the margin of stability the basic flow has a jet-like structure similar to that for $\Gamma^b = 0.975$. For $\Gamma^b = 0.8$ the jet is wider, however, and it separates from the wall further downstream. Thus the primary vortex extends down to $x \approx 7$. Hence, it is much more strained than for $\Gamma^b = 0.975$. Moreover, inertia effects displace the centre of the primary eddy downstream to $x \approx 5.5$.

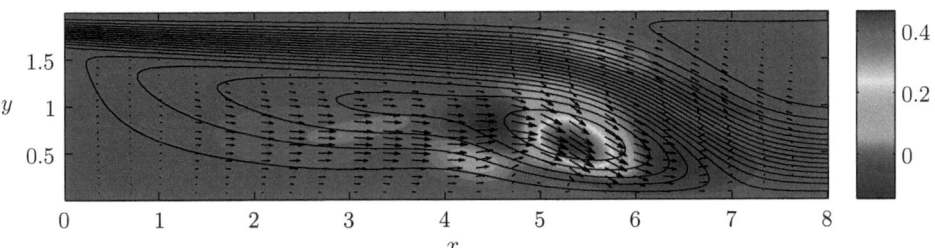

Figure 4.13.: Basic-flow streamlines for $\Gamma^b = 0.8$, critical velocity fields (arrows) and the total local energy production $\sum_i I_i$ for $\Gamma^b = 0.8$ in a plane in which it takes its global maximum.

The basic flow at criticality, the critical mode and the total local energy-transfer rate are shown in figure 4.13 in a cut, where the total local energy transfer takes its maximum value at $(x, y) = (5.3032, 0.5795)$. It is clearly seen that the energy transfer from the basic flow to the critical mode is entirely located in the core region of the primary eddy. Moreover, the perturbation flow is strongest in the core region of the primary vortex and aligned with the principal direction of strain.

In an (x, y) plane displaced by a quarter of the critical wavelength $\Delta z = \pi/(2k)$, the spanwise perturbation velocity takes its maximum amplitude, which is shown in figure 4.14. The two

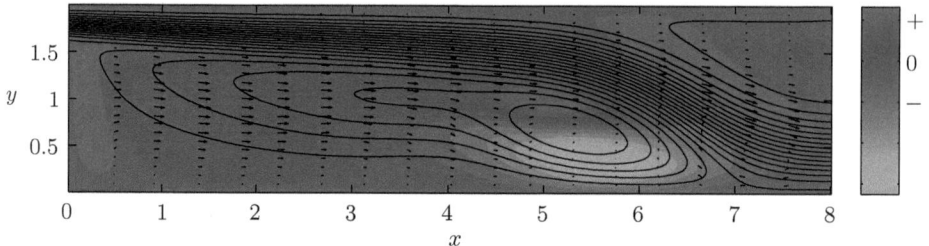

Figure 4.14.: Basic-flow streamlines for $\Gamma^b = 0.8$, critical velocity fields (arrows) and its spanwise component $\tilde{w}$ in a plane in which it takes its maximum amplitude.

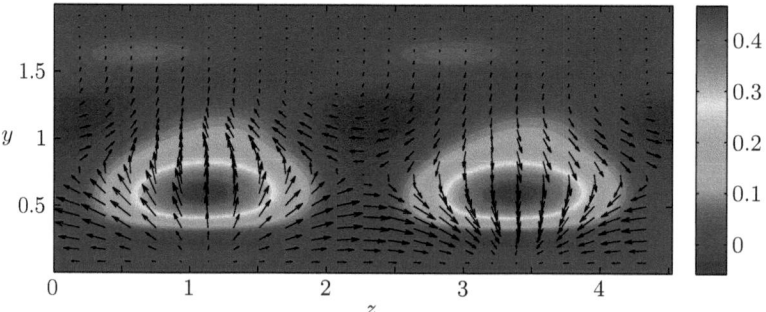

Figure 4.15.: Critical velocity fields (arrows) and the total local energy production $\sum_i I_i$ for $\Gamma^b = 0.8$ at $x = 5.3032$. The wave propagates to the left.

extrema of $\tilde{w}$ with different signs arise outside of the vortex core and they are aligned with the compressional strain axis. The resulting perturbation flow should arise as a sequence of counter-rotating vortices with vorticity perpendicular to the z axis and aligned with the dilatational strain axis. These vortices are not that clearly seen in figure 4.15, because the perturbation flow component $\tilde{u}$ strongly depends on x apparently resulting in sinks and sources in the projection of the perturbation flow onto a plane $x = \text{const}$. Figure 4.16 shows that the perturbation flow extends upstream up to the backward-facing step at $x = 0$, but no significant amplification is acting in that region.

The modal structure and the peak of the energy transfer in the centre of the strained primary vortex suggest that the instability is of elliptic type. Contrary to the usual stationary elliptic instability (see e.g. Kuhlmann et al., 1997; Pierrehumbert, 1986; Waleffe, 1990), the present instability is oscillatory (see e.g. Kerswell, 2002), which may be due to the asymmetric structure of the basic flow, which is not perfectly elliptic.

Both the centrifugal and the elliptical instability mechanisms are operative within the expansion ratio range $0.9 \lesssim \Gamma^b \lesssim 0.95$. The energy production associated with the two distinct

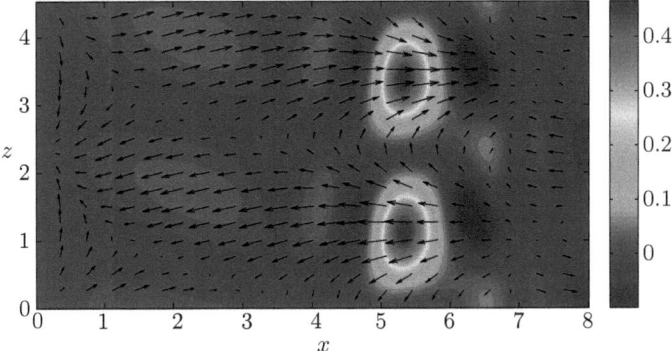

Figure 4.16.: Critical velocity fields (arrows) and the total local energy production $\sum_i I_i$ for $\Gamma^b = 0.8$ at $y = 0.5795$. The wave propagates from top to bottom.

local production extrema is of comparable magnitude for expansion ratios near $\Gamma^b \approx 0.94$ for which the critical frequency ω_c takes its maximum value (cf. figure 4.9). The elliptic instability mechanism is still dominating at the lowest expansion ratio $\Gamma - \Gamma^b_{co} \ll 1$ for which the critical mode is oscillatory.

Stationary instability for expansion ratios $\Gamma^b < \Gamma^b_{co}$

Instability near the codimension-two point $|\Gamma - \Gamma^b_{co}| \ll 1$ The stationary critical mode for $\Gamma^b \uparrow \Gamma^b_{co}$ and the oscillatory critical mode for $\Gamma^b \downarrow \Gamma^b_{co}$, and the respective energy-transfer rates, are very similar. This is illustrated in figure 4.17 for (a) the stationary ($\Gamma^b = 0.709 < \Gamma^b_{co}$) and ($b$) the oscillatory critical mode ($\Gamma^b = 0.7108 > \Gamma^b_{co}$), respectively. The differences are minute because the two critical modes are closely connected with each other in the parameter space (see figure 4.7).

In figures 4.17a and 4.17b the main energy transfer takes place within the centre of the primary vortex. The second weaker region of positive energy production is located between the primary and secondary vortices. This latter local maximum of I is more pronounced for the stationary mode at $\Gamma^b = 0.709$ and almost vanishes in the oscillatory case $\Gamma^b = 0.7108$. For I_2 the energy transfer takes place only in the centre of the primary vortex whereas I_1 is responsible for the second region of positive energy production. Both I_1 and I_2 are considered in planes where they take their local maxima.

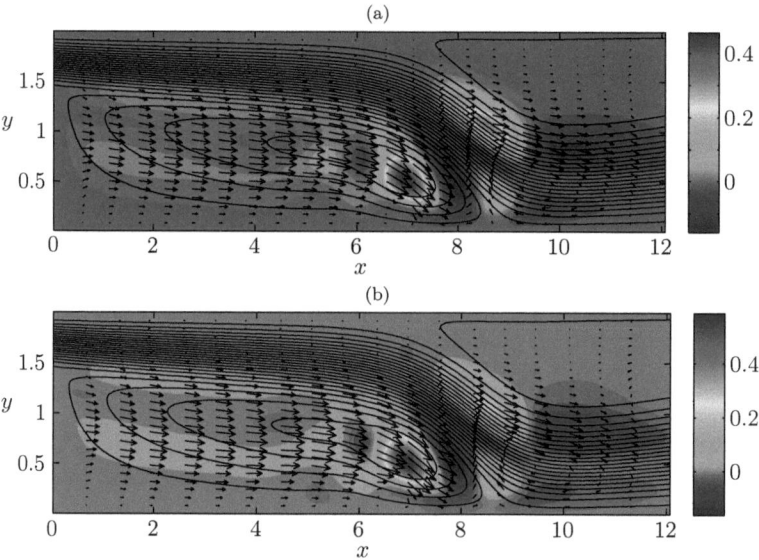

Figure 4.17.: Basic state (streamlines), critical velocity fields (arrows) and the total local energy production $\sum_i I_i$ in a z plane in which the total local production takes its absolute maximum: (a) stationary mode for for $\Gamma^b = 0.709 < \Gamma^b_{co}$; ($b$) oscillatory mode for $\Gamma^b = 0.7108 > \Gamma^b_{co}$.

Instability mechanisms for $0.4 \lesssim \Gamma^b < 0.709$ In the parameter range $0.4 \lesssim \Gamma^b < 0.709$ different mechanisms are present which makes the physical interpretation difficult. Throughout all computations the integral energy production $\int I'_i dV$ for $i = 1, 2, 3, 4$ are positive. Hence, all physical processes associated with these terms contribute to the destabilization process, albeit with different importance.

The most important terms in the range $0.4 \lesssim \Gamma^b < 0.709$ are I'_1 and I'_2. The term I'_1 describes an energy amplification by a cross-stream displacement $\tilde{\boldsymbol{u}}_\perp \cdot \nabla \boldsymbol{u}_0$ of the basic flow momentum. In order that this quantity feeds back on the perpendicular perturbation $\tilde{\boldsymbol{u}}_\perp$, the orientation of $\boldsymbol{u}_0$ must change perpendicular to itself. This implies that this energy term is associated with converging basic-flow streamlines. For $\Gamma^b = 0.5$ this is the case near the downstream end of the primary eddy just after the turning of the separated flow in the upstream direction (figure 4.18). The term I'_2 describes the usual lift-up mechanism by which basic flow momentum is transported in cross-stream direction ($\tilde{\boldsymbol{u}}_\perp \cdot \nabla \boldsymbol{u}_0$) and this feeds back on the parallel component of the perturbation flow $\tilde{\boldsymbol{u}}_\parallel$. The requirement here is a strong shear flow.

From the integral budgets in table 4.4 the two processes associated with I'_1 and I'_2 are the dominant contribution for $\Gamma^b \approx 0.7$. As Γ is further reduced, the process I'_4 becomes more and more important. It is the largest destabilizing factor in the integral energy budget for $\Gamma^b = 0.5$. The term I'_4 describes the streamwise transport of basic flow momentum $\tilde{\boldsymbol{u}}_\parallel \cdot \nabla \boldsymbol{u}_0$, which feeds back on the streamwise perturbation flow. The requirement for this process to be energy-producing is significant deceleration of the basic flow.

Figure 4.18 shonws the total local energy production (figure 4.18a) and the local production associated with the above-mentioned terms (figure 4.18b–d). All terms attain their local extrema in the same planes $z = \text{const}$. Even though the integral contribution of I'_4 dominates, the maximum production being located near the separation streamline where the flow is strongly decelerated upon approaching the reattachment point, this term does not leave its fingerprint in the total local budget, because the term I'_2 is locally strongly stabilizing in the same region (figs. 4.18c,d). The process I'_1 does not provide the largest integral contribution to the energy budget, but it exhibits the strongest local extrema, which is reflected in the total local production 4.18a.

To conclude, the flow instability for $\Gamma^b = 0.5$ is due to a combination of flow deceleration near the reattachment point (I'_4), a lift-up process (I'_2) on both sides of the jet (near the primary and the secondary eddy), and an amplification due to streamline convergence near and in the separated flow regions. The perturbation flow in the plane shown in figure 4.18 is nearly parallel to the basic flow, except for the region between the primary and the secondary eddies. Here, sizeable cross-stream perturbation-flow components exist, which can be clearly seen in the plane $x = 13.2728$ (location of the second maximum of the energy production rate) shown in figure 4.19.

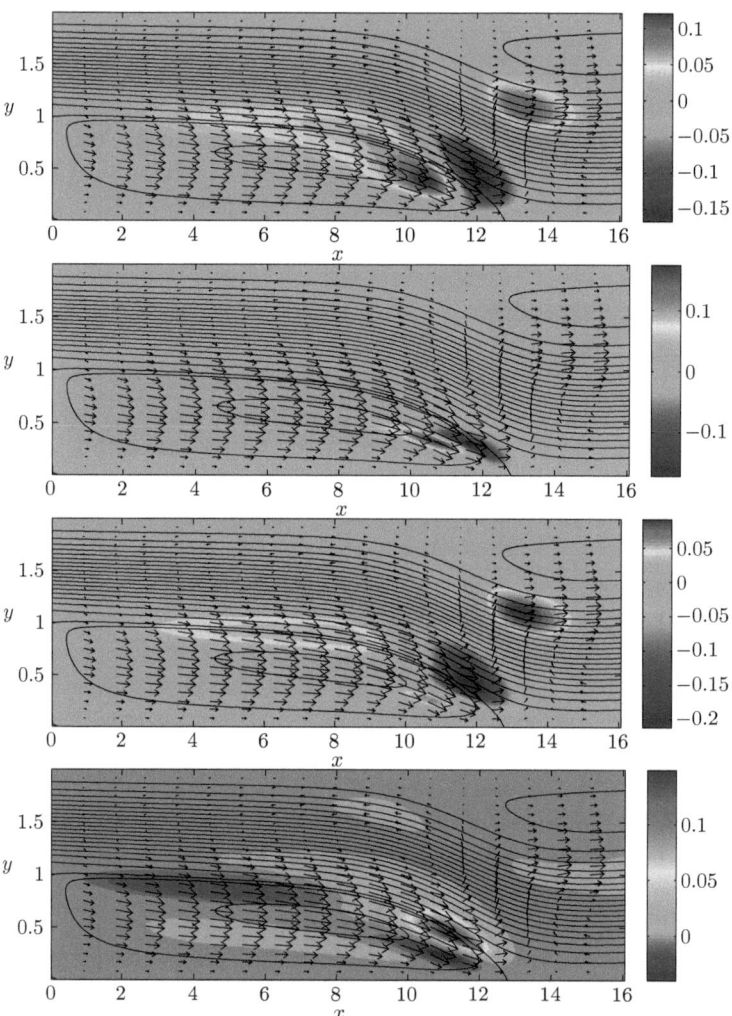

Figure 4.18.: Critical velocity fields (arrows) and local energy production rates for $\Gamma^{\mathrm{b}} = 0.5$ in a plane $z = \mathrm{const.}$ in which all production terms shown exhibit their local extrema: (a) total local production $\sum_i I_i$ for $\Gamma^{\mathrm{b}} = 0.5$ at $z = 0$, (b) I'_1, (c) I'_2 and (d) I'_4. The axes are not to scale.

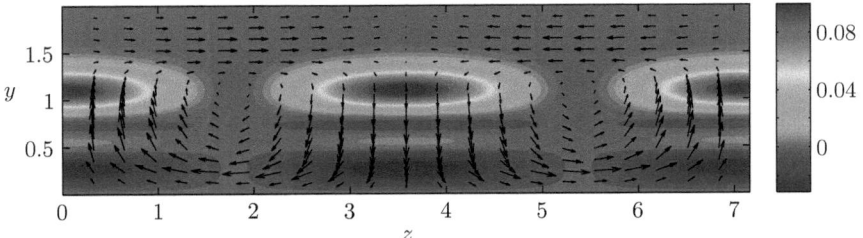

Figure 4.19.: Critical velocity fields (arrows) and the (total) local energy production $I'_2 \approx \sum_i I_i$ for $\Gamma^b = 0.5$ at $x = 13.2728$. The total local maximum energy transfer, however, takes place at $x = 10.5662$.

Instability by lift-up and flow acceleration for $0.25 \leq \Gamma^b \lesssim 0.4$ When the expansion ratio is small the Reynolds number must be relatively high in order to destabilize the flow. In this situation the primary recirculation zone of the basic flow becomes strongly elongated in the x direction. This can be seen from figure 4.20, where the primary vortex of the basic flow is extremely stretched (note the scaling of the x and y axes). For that reason the local basic flow at a downstream position is almost parallel. Thus one might expect some kind of Kelvin–Helmholtz instability. This is, however, not observed. The reason is that, despite the elongation of the primary vortex, the flow behind the step is characterized by the recirculation. The residence time of a fluid element in the pure shear layer that forms downstream from the step is too short to enable a significant growth of typical Kelvin–Helmholtz vortices. Besides of that, the Kelvin–Helmholtz instability is two-dimensional in nature.

A different instability is found, which is steady and three-dimensional from the outset. For $\Gamma^b = 0.25$ the critical mode at $\text{Re}_c = 4912.8$ with $k_c = 1.2477$ is confined to the region of the primary vortex. From figure 4.20 it is seen that the perturbation flow is amplified in the shear layer around the separating streamline.

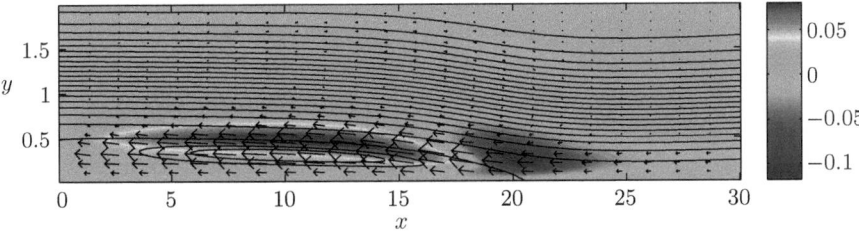

Figure 4.20.: Critical velocity fields (arrows) and the total local energy production $\sum_i I_i$ for $\Gamma^b = 0.25$ at $z = 0$.

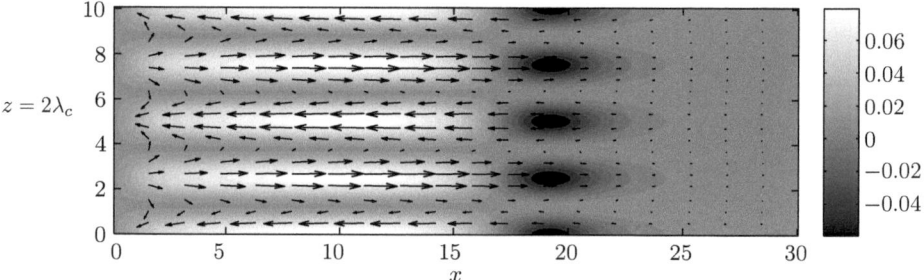

Figure 4.21.: Critical velocity fields in the form of streaks (arrows) and total local energy production $\sum_i I_i$ for $\Gamma^b = 0.25$ at $y = 0.4737$.

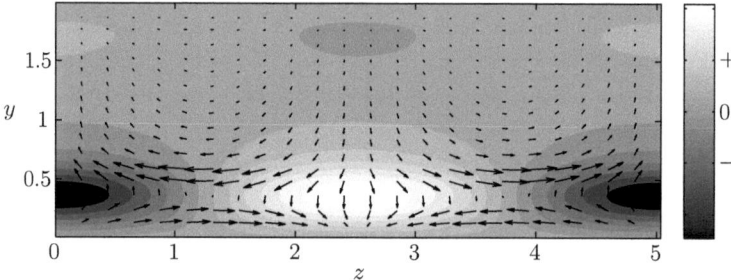

Figure 4.22.: Critical velocity fields (arrows) and the streamwise perturbation velocity $\tilde{u}$ (grey shading) for $\Gamma^b = 0.25$ at $x = 10.4632$. As in figure 4.15 the streamwise vortices are somewhat obscured due to the strong variation $\partial_x \tilde{u}$.

The critical mode takes the form of alternating slow and fast streaks, which are shown in figure 4.21. Figure 4.22 shows a cross section through the streaks and the connecting lateral flow in the z direction.

It is well known that such streaks are produced by the lift-up mechanism. Landahl (1980) showed that a wide class of localized initial three-dimensional disturbances evolve into longitudinal streaks for any inviscid shear flow. This streaky structure is physically explained by the lift-up mechanism where fluid with low velocity is lifted up from the wall and interacts with high-speed regions (Landahl (1975)). This leads to low- and high-speed streaks alternating in the spanwise direction. Different from open parallel flows, these streaks are self-sustained here owing to the feedback provided by the primary eddy recirculation. From figure 4.20 and 4.21 it can be seen that the streamwise perturbation component $\tilde{u}$ clearly dominates $\tilde{v}$ and $\tilde{w}$. The lift-up effect creates large streamwise perturbations via transport of the base-flow momentum by cross-stream velocity perturbations $-\tilde{v}\partial_y u_0$. In fact, this product is part of I_2 which is by far the most dominant term for $\Gamma^b = 0.25$.

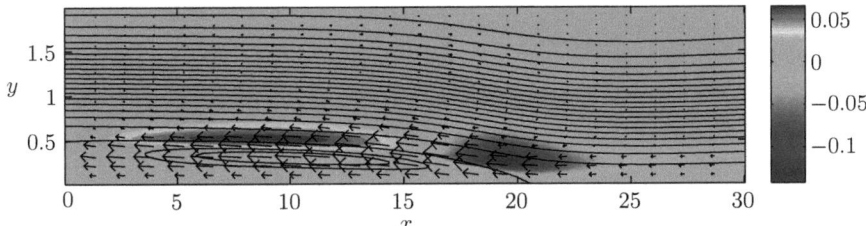

Figure 4.23.: Critical velocity fields (arrows) and the local energy production I'_2 for $\Gamma^b = 0.25$ at $z = 0$.

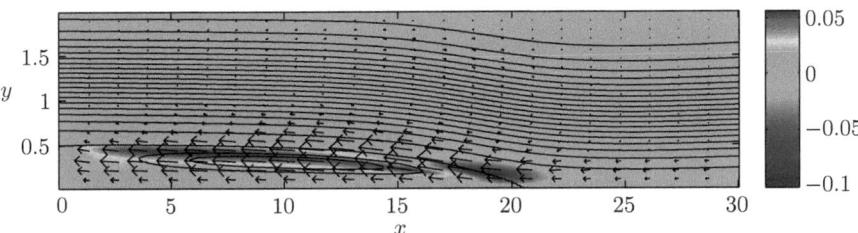

Figure 4.24.: Critical velocity fields (arrows) and the local energy production I'_1 for $\Gamma^b = 0.25$ at $z = 0$.

The most dominant term in the energy production for the streamline decomposition is I'_2 (figure 4.23). This term describes the classical lift-up mechanism and acts just above the separating streamline. The energy transfer term I'_1 (shown in figure 4.24) also contributes substantially to the destabilization. This term describes an energy transfer due to basic-flow convergence. Such conditions are met near the upstream end of the primary eddy where the flow accelerates the streamlines being non-parallel and slightly converging.

To conclude, the instability in the expansion ratio range $0.25 \leq \Gamma^b \lesssim 0.4$ is due to the lift-up mechanism and streamline convergence of the flow near the separating streamline of the primary eddy recirculation, which provides a feedback for the perturbations to be self-sustained.

Comparison with experimental results

It is desirable to compare the critical data and the numerically obtained basic and perturbation flows with experimental results. Such a comparison is made difficult for several reasons. In some experiments the spanwise aspect ratio $\Lambda = L_z/H$ where L_z is the width of the channel, has been selected to be very small such that the no-slip conditions on the side walls induce a significant three-dimensionality of the flow for Reynolds numbers much lower than the critical ones. Other experiments with relatively large spanwise aspect ratio were not aimed at a precise

Γ^b	Re	authors	x_r^l	x_s^u	x_r^u
0.4851	600	Armaly et al. (1983) (exp.)	14	11.4	20.0
		present	11.73	9.25	20.0
0.5	603.75	Lee & Mateescu (1998) (exp.)	12.9	10.3	20.5
		present	11.82	9.41	20.8
0.5	600	Erturk (2008)	11.83	9.48	20.55
		Mateescu & Venditti (2001)	11.8	9.32	20.62
		present	11.81	9.4	20.53

Table 4.5.: Comparison of separation and reattachment positions on the upper (x_s^u, x_r^u) and lower wall (x_r^l) with previous experimental and numerical results.

determination of the critical conditions and the structure of the flow at onset. Nevertheless, a comparison is made here with the available experimental data.

Beaudoin et al. (2004) experimentally studied the flow over a backward-facing step in a geometry with small expansion ratios $\Gamma^b = 0.05$ and $\Gamma^b = 0.1$, which are not covered in the present study. The spanwise aspect ratio of their system was only $\Lambda = 1.5$. Hence, they observed three-dimensional flows at rather low Reynolds numbers, an effect they traced back to strong sidewall effects. Clearly, such sidewall effects would obscure any of the bulk flow instabilities considered above. Nevertheless, the stationary, counter-rotating longitudinal vortices observed experimentally are qualitatively similar to our findings for small expansion ratios.

The experiments of Armaly et al. (1983) ($\Lambda = 17.82$) and Lee & Mateescu (1998) ($\Lambda = 20$) were devoted to the study of two-dimensional flows over a backward-facing step. Therefore, much larger spanwise aspect ratios were realized in these investigations in order to minimize three-dimensional sidewall-induced flows. Table 4.5 shows measured separation and reattachment positions on the upper and lower walls in comparison with two-dimensional computations. The data match very well. Also the simulations of Erturk (2008) and Mateescu & Venditti (2001) are shown, since these authors used an inlet channel upstream of the step, as it was done here.

Laser Doppler measurements of Armaly et al. (1983) were carried out using an experimental setup with expansion ratio $\Gamma^b = 0.4851$. They found a sizeable deviation from a pure two-dimensional flow for Reynolds numbers exceeding Re $\approx$ 300. Williams & Baker (1997) performed three-dimensional numerical simulations for a geometry exactly matching the nominal geometry of the apparatus used by Armaly et al. (1983). By Lagrangian particle tracking it was demonstrated that the lateral sidewalls were responsible for the three-dimensionality of the flow when Re $\geq$ 300. Apart from these sidewall-induced perturbations of the two-dimensional flow, Armaly et al. (1983) observed that the laminar flow regime extends up to Re = 874 (according to the definition of Re). Beyond this value the locations of the measured detachment and reattachment lines feature a sharp kink. This jump has been identified as an indication for the onset of the transitional Reynolds number range. This observation is in accordance with our critical Reynolds number $\text{Re}_c(\Gamma^b = 0.4851) = 768.74$. It can thus be concluded that the marked changes of the detachment and reattachment lines are due to the bulk-flow instability. The findings of Armaly et al. (1983) confirm that the properties of the experiments are fully consistent with our linear stability analysis. According to Armaly et al. (1983) fully developed turbulent flows arise only for $\text{Re} \geq 4807$.

Hot-film sensor measurements were presented by Lee & Mateescu (1998) for $\text{Re} \leq 2250$ and an expansion ratio $\Gamma^b = 0.5$. Their results agree with those of Armaly et al. (1983) for an expansion ratio $\Gamma^b = 0.4851$ with a typical discrepancy of about 8 %. Lee & Mateescu (1998) reported that the transitional flow regime starts at $\text{Re} \geq 862.5$, which is consistent with our calculations yielding $\text{Re}_c(\Gamma^b = 0.5) = 714.05$.

The systematic deviation of the observed loss of stability of the two-dimensional flow in the experiments from our computed critical Reynolds numbers, which are lower by about 14% and 20%, is most likely due to a stabilizing effect of the sidewalls. This stabilization may be due to the flow or to the fact that the optimum (critical) wavelength is not an integer multiple of the spanwise aspect ratio. Another possible reason could be the finite amplitude that is required for a measurable signal in the case of a forward bifurcation.

In summary, the present modal stability boundaries qualitatively fit the experimental results. In contrast, the critical Reynolds number $\text{Re}_c^B(\Gamma^b = 0.5) = 57.7$ determined by Blackburn et al. (2008) via a transient-growth analysis is one order of magnitude smaller.

4.1.4. Conclusion

The global temporal linear stability of the two-dimensional flow over a backward-facing step has been investigated numerically. The inlet and outlet channel lengths have been carefully selected to ensure channel-length-independent results. Stability boundaries have been computed for a quasi-continuous variation of the channel expansion ratio Γ^b. An *a posteriori* energy-transfer analysis revealed that the flow becomes unstable to three-dimensional perturbations due to different mechanisms.

Even though the basic flow at criticality and for very large expansion ratios $(1 - \Gamma) \ll 1$ resembles a wall jet immediately behind the step, the typical two-dimensional wall-jet instability was not found (see e.g. Chun & Schwarz, 1967). The reason is that the jet emerging from the inlet separates from the top wall, turns downwards and impinges nearly perpendicularly on the bottom wall. Most of the kinetic perturbation energy is produced on the downstream side of the jet in a region in which centrifugal forces are significant. The spatial structure of the critical mode, which is oscillatory, is similar to the one found in deep one-sided lid-driven cavities (see figure 20 of Albensoeder et al., 2001) where all instabilities are of centrifugal type.

When the step height is decreased moderately, the separation of the jet from the upper wall is delayed and the primary separated vortex becomes increasingly strained at the critical Reynolds number. The oscillatory critical mode and the energy-transfer characteristics change continuously until, at $\Gamma^b \approx 0.8$, the entire perturbation-energy production takes place in the centre of the elliptically strained primary vortex. This is also the locus of highest perturbation-flow amplitude. The energy-transfer characteristics and the structure of the critical mode suggest that the instability is of elliptic type. The nature of the instability does not change on a further decrease of Γ^b down to $\Gamma^b_{co} \approx 0.71$ where the critical mode changes from oscillatory to stationary. Even though the temporal character of the instability changes at Γ^b_{co} the respective energy production rates are very similar to each other in magnitude and spatial distribution, since the two critical modes are closely connected with each other in the parameter space.

As the step height is further reduced the primary vortex at criticality becomes even more strained and its extension grows downstream. The perturbation energy production pattern now features two well-separated and pronounced maxima. It is argued that the instability cannot be attributed to any classical instability mechanism. Rather, the flow instability is caused by a combined effect of flow deceleration near the reattachment point of the separating streamline of the primary vortex, a lift-up process on both sides of the jet emerging from the step, and streamline convergence in the regions of separated flow. No indications were found for a centrifugal instability for $\Gamma^b = 0.5$ as proposed by Barkley et al. (2002).

For small step heights ($0.25 \leq \Gamma^b \lesssim 0.4$) the separated vortex is strained even more and the basic flow at criticality becomes nearly parallel. In this situation the perturbation amplification is entirely located near the separating streamline within the shear layer. Moreover, the critical mode exhibits alternating slow and fast streaks connected by streamwise vortices in and near the separated primary vortex. The energy-transfer analysis shows that the perturbation streaks are fed by the lift-up mechanism. Other than in plane Couette and Poiseuille flow, the streaks and streamwise vortices are self-sustained, in the present case, because of the feedback provided by the primary vortex. The lift-up mechanism is supported by a *cross-stream* momentum-transfer mechanism $\tilde{\boldsymbol{u}}_\perp \cdot \nabla \boldsymbol{u}_0$, which yields an energy growth (I'_1) in the region of converging basic flow near the upstream end of the primary vortex.

The critical data computed are consistent with previous experimental findings. Therefore, the physical relevance of the global stability modes is confirmed. The prevailing results of the global linear stability analysis will thus provide useful reference data for further investigations. It was shown that previous experiments suffered from a too small spanwise aspect ratio. Thus a distinction between sidewall effects and bulk-flow instabilities is made difficult. Hence, it would be very interesting to study more systematically finite-size effects, caused by the presence of rigid walls that limit the spanwise extent of the system. Of interest are the structures of pure sidewall-induced flows and the influence of these on the stability boundaries and associated modifications of the instability mechanism.

4.2. The Forward-Facing-Step Problem

4.2.1. Problem Definition

The analysis is concerned with an incompressible flow of a Newtonian fluid over a rectangular forward-facing step in the (x, y) plane. The flow domain is assumed to be infinitely extended in the spanwise (z) direction. The geometry is depicted in figure 4.25, consisting of an inlet channel of height H and length L_i followed by a suddenly constricted channel of height h_o and length L_o. The origin of the coordinate system is located at the bottom of the step of height $h_s = H - h_o$.

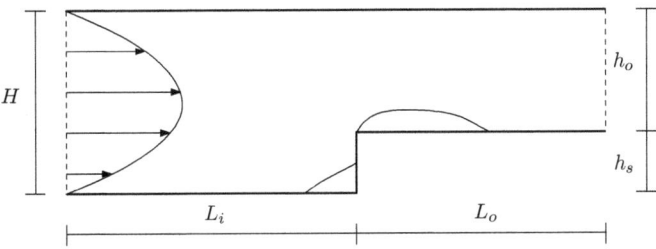

Figure 4.25.: Sketch of the flow domain for the forward-facing step problem.

Here, the ideal case is considered, in which both L_i and L_o tend to infinity. For the numerical treatment, however, both channels must be truncated. The corresponding finite lengths will be selected sufficiently large such that the flow near the step is independent of L_i and L_o. The problem is then governed only by two parameters: the Reynolds number Re and the constriction ratio Γ^{f} defined as

$$\mathrm{Re} = \frac{H \bar{U}}{\nu} \quad \text{and} \quad \Gamma^{\mathrm{f}} = \frac{h_s}{H}. \quad (4.2.1)$$

Here the mean velocity $\bar{U}$ of the inlet flow, the height H of the inlet channel and the kinematic viscosity ν of a Newtonian fluid have been used for the scaling.

At the inlet $x = -L_i/H$, plane Poiseuille flow $u_0(y) = 6y(1-y)$ and $v_0 = 0$ is assumed for $y \in [0; 1]$. Along the step and all channel walls, no-slip and no-penetration boundary conditions $\boldsymbol{u}_0 = 0$ are adopted. At the outlet $x = L_o/H$, homogeneous Neumann conditions $\partial_x \boldsymbol{u}_0 = 0$ are used for all velocity components, and the pressure level is specified by $p_0 = 0$.

4.2.2. Scientific Background

The flow over a forward-facing step in a plane channel is a prototype example for flow separation at a sharp edge near the front of a bluff body. The forward-facing-step configuration often represents the entrance geometry for polymer processes like fibre spinning, film blowing and extrusion (Chiba et al., 1995) and is important for the design of heat-transfer devices, such as combustion chambers, cooling systems and heat exchangers. Apart from its technical applications, forward-facing-step flows are also observed in rivers and lakes and might be responsible for erosional grooves (Pollard et al., 1996).

Earlier publications (see e.g. Dennis & Smith, 1980; Plotkin & Mei, 1986) on the forward-facing-step flow focused on the pure two-dimensional flow and on the dependence of the length of the separation zone on the Reynolds number. The topology of the two-dimensional flow is characterized by two separation zones: a first recirculation bubble ahead of the step even for low Reynolds numbers and, for larger Reynolds numbers, a second recirculation zone arises past the step which separates immediately at the sharp corner.

In many experiments (Chiba et al., 1995; Pollard et al., 1996; Stüer, 1999; Stüer et al., 1999), it was reported that the streamlines in front of the step are not closed but rather open and three-dimensional in nature which enables an entrainment into the recirculation bubble. This three-dimensionality, giving rise to the formation of streaky structures downstream of the step corner, was observed even at low Reynolds numbers. The work of Stüer et al. (1999) considered mainly the issue of whether the separation bubble in front of the step is closed or open by studying the local three-dimensionality, but did not pay particular attention to the matter when the flow gets unstable and consequently globally three-dimensional. Nevertheless, the hydrogen bubble visualization technique revealed that the flow is turbulent at a Reynolds number (according to (4.2.1)) of around 4800.

Wilhelm et al. (2003) revealed in their numerical simulations that the three-dimensionality in front of the step is a sensitive reaction of the flow to minute disturbances present in the oncoming flow. Perturbation amplitudes of less than 1 % of the mean flow, which are most likely present in the laboratory, can grow to sizeable amplitudes just ahead of the step. Moreover, it was demonstrated that there exists no significant difference between the two-dimensional and spatially averaged three-dimensional flow fields. Recently, Marino & Luchini (2009) confirmed via an adjoint analysis the findings of Wilhelm et al. (2003) that the flow is highly sensitive to oncoming disturbances and showed that this significant receptivity is independent of the presence of a hydrodynamic instability. Therefore, the three-dimensionality as observed in the experiments cannot be attributed to a global instability.

Stüer (1999) investigated also numerically the flow over a forward-facing step with constriction ratios of $\Gamma^f = 0.5$ and $\Gamma^f = 0.25$. By a global linear stability analysis, it was shown that the critical Reynolds numbers are $\mathrm{Re}_c(\Gamma^f = 0.5) \approx 150$ and $\mathrm{Re}_c(\Gamma^f = 0.25) \in [300; 400]$. The latter result is in contradiction with the findings of Wilhelm (2000) and Wilhelm et al. (2003), who showed that the flow for $\Gamma^f = 0.25$ subjected to three-dimensional perturbations is absolutely stable for Re = 1320, representing a lower bound to the stability boundary. Recently, Marino & Luchini (2009) computed the stability boundaries for the constriction ratios

$\Gamma^f = 0.5$ and $\Gamma^f = 0.25$. They obtained critical Reynolds numbers of $\text{Re}_c(\Gamma^f = 0.5) \approx 997$ and $\text{Re}_c(\Gamma^f = 0.25) \approx 3440$, which are approximately 10 times higher than the ones of Stüer (1999).

The present study is conducted on the one hand to clarify the discrepancy between the critical Reynolds numbers of Stüer (1999) and Marino & Luchini (2009) and on the other hand to understand the physical nature of the instability, which has never been examined for the forward-facing-step problem. Hence, a global stability analysis of the two-dimensional basic flow is carried out with a systematic variation of the geometry, covering a wide range of constriction ratios from $\Gamma^f = 0.23$ to $\Gamma^f = 0.965$. It will be shown that the critical data obtained by Marino & Luchini (2009) are not fully grid independent and clearly depend on the length of the inlet channel. The grid- and entrance-length-independent critical Reynolds numbers are up to 1.7 times larger than those obtained by Marino & Luchini (2009). In addition, the instability mechanism of the global modes will be elucidated by analysing the kinetic energy transferred from the basic flow to the critical mode. The prevailing work also comprises a structural sensitivity analysis in order to assess the results of the linear stability analysis.

4.2.3. Results

Code verification

The critical data for the forward-facing-step flow are computed for the same geometric parameters as Marino & Luchini (2009). Using $L_i = 1H$ and $L_o(\Gamma^f = 0.5) = 6h_o$ and $L_o(\Gamma^f = 0.25) = 5.\dot{3}h_o$, respectively, $\text{Re}_c = 1037.8\,(997)$ and $k_c = 6.0975\,(6.2)$ are obtained for $\Gamma^f = 0.5$ and $\text{Re}_c = 3469.7\,(3440)$ and $k_c = 5.9124\,(6.4)$ for $\Gamma^f = 0.25$, where the values of Marino & Luchini (2009) are given in parentheses. Both sets of data agree well. However, the results of the above-mentioned authors are not fully grid converged.

Grid convergence was obtained by using the minimum cell size of $\Delta x_{\min} = \Delta y_{\min} = 10^{-3}$ (in units of H) along the walls and around the step. The cell width is increased smoothly using a stretching factor of 1.03 until the maximum grid spacings $\Delta x_{\max} = 0.02$ and $\Delta y_{\max} = 0.01$ are reached. The number of grid points used are shown in table 4.7 for representative constriction ratios. N_x^i and N_y^i represent the number of grid points in the x and y directions, respectively, in the inflow channel, while N_x^o and N_y^o are the number of grid points in the outflow channel.

As a final verification, the pressure distribution is compared to the one obtained by ANSYS FLUENT 13.0. The grid is built with the same number of grid points and the same grid resolution where a hyperbolic tangent stretching function is used to distribute the grid points. The pressure and the gradients are resolved with methods of second order and for the convective terms the QUICK scheme is applied. In figure 4.26 the pressure distribution of the two-dimensional basic state p_0 is shown slightly above the step at $y = 0.2505$ for $\text{Re} = 1320$ and $\Gamma^f = 0.25$. Dashed lines represent the present solution and dotted ones the result of ANSYS FLUENT 13.0. Both curves are indistinguishable on the scale shown. The pressure has steep

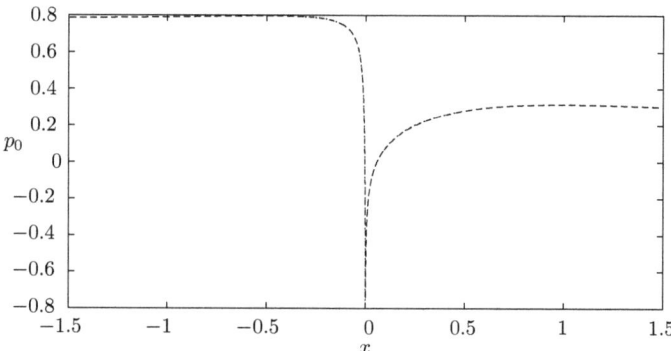

Figure 4.26.: Pressure distribution of the basic state at $y = 0.2505$ for Re = 1320 in the vicinity of the step corner for $\Gamma^f = 0.25$. Dashed lines depict the present result and dotted curves the solution of ANSYS FLUENT 13.0.

gradients in the vicinity of the step due to the corner singularity and features a minimum of -0.7532 at $x = 0.0005$ with the present code. ANSYS FLUENT 13.0 yields a minimum of -0.7519, where the pressure level is set to zero at the outlet.

Parameter dependence

To obtain critical Reynolds and wave numbers for infinitely long in- and outlet channels, the influence of L_i and L_o on Re_c and k_c was studied. The length of the entrance channel L_i turns out to be of utmost importance. If the inflow channel is too short, the basic flow in the vicinity of the step is not fully developed to the basic flow for $L_i \to \infty$. This is shown in figure 4.27, where the basic flow is depicted for $L_i = 8H$ and $L_i = 1H$ for Re = 3518.4 and $\Gamma^f = 0.25$, using the same outflow-channel length $L_o = 32h_o$. It can be seen that for $L_i = 1H$ the length of the first separation bubble is too short and the one of the secondary vortex is too long.

The entrance-channel-length effect on the stability boundaries is shown in table 4.6. Results are given for $\Gamma^f \in \{0.25, 0.5, 0.75\}$, but a similar convergence of the critical data, as L_i is increased, is obtained for all constriction ratios covered here. As can be seen from table 4.6, the length of the inlet channel should be at least $L_i = 6H$ to restrict the inlet-channel-length effect on Re_c to below $\approx 0.03\%$.

The influence of the outlet-channel length L_o on the stability boundaries was also scrutinized (not shown here). It is found that $L_o = 32h_o$ was more than sufficient to keep the effect on Re_c and k_c below $\approx 0.01\%$. Figure 4.28 shows for $\Gamma^f = 0.5$, Re = 1423.9 and $L_i = 8H$ the basic velocity profiles $u_0(y)$ at the outlet for two different outflow channel lengths, $L_o = 32h_o$ and $L_o = 6h_o$ (the choice of Marino & Luchini, 2009). It can be noticed that the velocity profile for

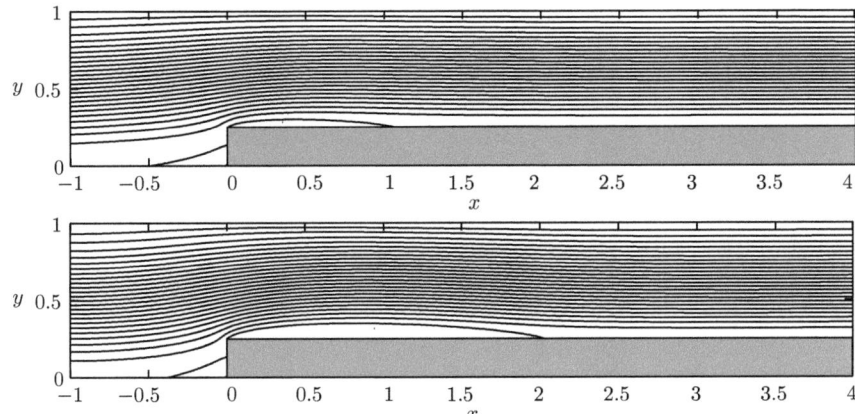

Figure 4.27.: Streamlines of the base flow for $\Gamma^f = 0.25$, Re = 3518.4 and $L_o = 32h_o$ for different inflow channel lengths: (a) $L_i = 8H$ and (b) $L_i = 1H$.

L_i	1H	2H	3H	6H	8H	12H	Γ^f
Re_c	3518.4	5290.6	5741.9	5886.7	5888.4	5888.4	0.25
k_c	5.9424	7.2459	7.6341	7.7601	7.7621	7.7622	0.25
Re_c	1048.1	1366.5	1413.1	1423.8	1423.9	1423.9	0.5
k_c	6.1240	6.7581	6.8615	6.8831	6.8833	6.8833	0.5
Re_c	771.89	843.16	846.89	848.50	848.51	848.51	0.75
k_c	11.075	11.459	11.4998	11.507	11.508	11.508	0.75

Table 4.6.: Critical values Re_c and k_c as functions of the entrance length L_i for multiples of the characteristic length scale H for three different constriction ratios and $L_o = 32h_o$.

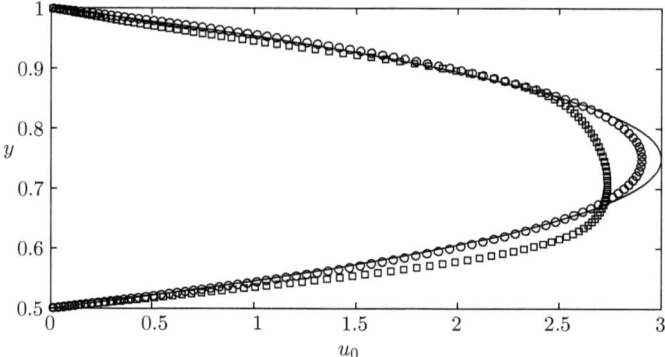

Figure 4.28.: Velocity profiles $u_0(y)$ at the outlet for $\Gamma^{\mathrm{f}} = 0.5$, Re = 1423.9 and $L_i = 8H$ shown for $L_o = 6h_o$ (squares) and for $L_o = 32h_o$ (dots). The solid line represents the fully developed plane Poiseuille flow.

$L_o = 6h_o$ deviates substantially from the fully developed plane Poiseuille flow; it is not even symmetric with respect to the mid-channel height.

To sum up, all results reported in the present study are computed by using an entrance length of $L_i = 8H$ and an outflow length of $L_o = 32h_o$ to minimize deviations from the asymptotic limit $(L_i, L_o) \to \infty$. For this configuration and using more than twice the number of grid points in the y direction as Marino & Luchini (2009), stability boundaries up to 70 % higher than the ones of these authors are obtained.

Stability boundaries

Representative numerical critical Reynolds and wave numbers are collected in table 4.7 for selected expansion ratios. Also, the number of grid points used is indicated.

Figure 4.29 shows critical values as functions of the constriction ratio Γ^{f}. The stability boundary varies smoothly with the constriction ratio. All critical modes are stationary. The smallest constriction ratio considered was $\Gamma^{\mathrm{f}} = 0.23$ with a critical Reynolds number of $\mathrm{Re}_c = 7135$. This Reynolds number is below the linear stability boundary of the plane Poiseuille flow, which is $\mathrm{Re}_c^{\mathrm{PPF}} \approx 7700$ for the present scaling. For larger constriction ratios (increasing step heights) Re_c decreases continuously and seems to approach an asymptotic limit $\mathrm{Re}_c \approx 680$ for $\Gamma^{\mathrm{f}} \to 1$.

In the limit $\Gamma^{\mathrm{f}} \to 1$, the height of the outlet channel h_o remains the only geometrical length scale. The correspondingly defined Reynolds and wave numbers are, respectively,

$$\mathrm{Re}^* = \mathrm{Re}\frac{h_o}{H}, \qquad k^* = k\frac{h_o}{H}, \qquad \text{for } \Gamma^{\mathrm{f}} \in [0.5; \to 1[. \qquad (4.2.2)$$

Γ^f	Re_c	k_c	N_y^i	N_y^o	N_x^o
0.23	7135.0	8.2324	276	174	1314
0.25	5888.4	7.7621	278	172	1282
0.3	3902.6	7.0462	282	166	1201
0.4	2135.4	6.6002	288	156	1039
0.5	1423.9	6.8833	292	146	878
0.6	1083.3	7.8358	288	132	716
0.7	904.41	9.8126	282	116	554
0.8	807.38	14.120	270	94	393
0.9	751.02	27.434	248	62	231
0.95	706.21	53.838	228	38	150

Table 4.7.: Critical data for selected constriction ratios Γ^f. The number of grid points (N_x, N_y) in the (x, y) plane is also given. For all calculations $N_x^i = 473$ is used.

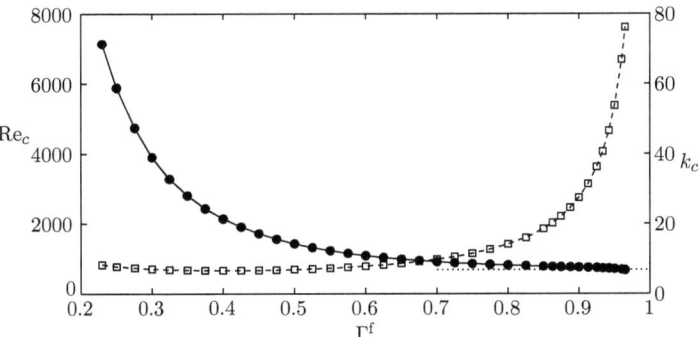

Figure 4.29.: Critical Reynolds number Re_c (dots and full line) and wave number k_c (squares and dashed line) as functions of the constriction ratio Γ^f. The dotted line represents the extrapolated value for $\Gamma^f \to 1$ with $Re_c \approx 680$.

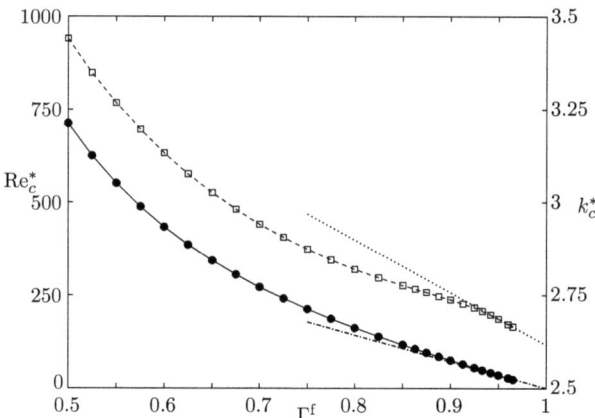

Figure 4.30.: Critical Reynolds number Re_c^* (dots and full line) and wave number k_c^* (squares and dashed line) as functions of the constriction ratio Γ^f. The data are scaled according to (4.2.2). The dash-dotted and dotted lines represent the best estimate for an asymptotic behaviour for $\Gamma^f \to 1$ as mentioned in the text.

Using the above scaling, the critical data are depicted in figure 4.30. The critical Reynolds number suggests a linear asymptotic scaling as $\Gamma^f \to 1$. The linear scaling is estimated by a linear fit through the three discrete data points for the largest values of Γ^f, resulting in $\mathrm{Re}_c^* \approx 711.94(1 - \Gamma^f)$. A linear scaling of k_c^* for $\Gamma^f \to 1$ is not as obvious. Nevertheless, the dotted line in figure 4.30, being obtained in the same way as for Re_c^*, corresponds to $k_c^* \approx 4.0272 - 1.4115\Gamma^f$.

Regardless of the constriction ratio Γ^f, the two-dimensional basic flow over a forward-facing step features a region of separated flow in the 90° corner in front of the step. This vortex is called the primary vortex, as it is the first one to appear as Re is increased. For higher Reynolds numbers, a secondary recirculation bubble arises which separates immediately downstream of the sharp corner of the step. This secondary vortex is of key importance for the flow instability, as the critical modes and their associated energy-production rates are most pronounced near this region. Therefore, the flow characteristics shall be discussed near the secondary vortex, although the computational domain is much more extended in the x direction.

Instability by lift-up and flow deceleration

The analysis shows that the physical instability mechanism is independent of the constriction ratio. As a representative case, the instability for $\Gamma^f = 0.25$ is discussed, because this constriction ratio was also analysed by Wilhelm et al. (2003). For this purpose, it proved useful to compute the kinetic energy-transfer rates between the basic and the perturbation flows. The

i	1	2	3	4	$\int_{S_o} I_5 \mathrm{d}S$
$\int_V I'_i \mathrm{d}V$	0.0508	0.2166	0.0869	0.7246	−0.0788
$\int_V |I'_i| \mathrm{d}V$	0.1211	0.5404	0.0879	0.2506	

Table 4.8.: Global normalized energy-production rates for $\Gamma^{\mathrm{f}} = 0.25$.

normalized energy-production rates in streamline coordinates I' integrated over the whole flow domain are summarized in table 4.8 for $\Gamma^{\mathrm{f}} = 0.25$. Note that all normalized energy-production terms sum up to unity with accuracy better than 10^{-4}.

The critical mode at $Re_c = 5888.4$ and $k_c = 7.7621$ is confined to the region of the secondary separation bubble, which arises for Re $\gtrsim 770$ if $\Gamma^{\mathrm{f}} = 0.25$. From figure 4.31a, it can be seen that the disturbances gain most of the energy in the shear layer around the separating streamline.

The critical mode appears as slow and fast streaks alternating in the spanwise direction, which are shown in figure 4.32 for two periods $2\lambda_c = 4\pi/k_c$. The critical wavelength is approximately three times the step height, which agrees well with the findings of Wilhelm et al. (2003) and the observations in the experiments of Stüer et al. (1999).

The streaky structure can be explained by the classical lift-up mechanism (Landahl, 1975, 1980). As already mentioned in §4.1.3, this effect generates large-amplitude velocity perturbations in the streamwise direction by the weak lift-up of fluid particles in wall-normal direction in the presence of a strong shear flow. Figures 4.31a–d indicate that the streamwise perturbation velocity $\tilde{u}$ prevails over the wall-normal component $\tilde{v}$. The streamwise vortices creating the streaks, are clearly localized in the immediate vicinity of the step. This is confirmed in figure 4.33, which shows the maximum over y of the absolute amplitudes $|\hat{v}(x,y)|$ and $|\hat{w}(x,y)|$. The amplitudes are monotonically decaying in the streamwise direction, which is shown in the inset of figure 4.33 for $x \in [3; 24]$.

The streamwise streaks, however, do not decay in a monotonic way like the streamwise vortices. Figure 4.34 shows the modulus of $|\hat{u}(x,y)|$ as function of the streamwise coordinate x. It can be seen that the amplitudes of the streaks start to increase at around $x \approx 3.6$, where the sign of $\hat{u}$ changes as function of x (the direction of the streaks is revised). The strength of the streaks is compatible with the strength of the weak streamwise vortices. As the streamwise vortices seem to decay, it is expected that also the streaks will decay ultimately further downstream (beyond the present computing power).

Schmid & Henningson (2001) demonstrated that the lift-up mechanism can give rise to transient growth for parallel shear flows. By contrast, the disturbances are localized near the step and self-sustained here on account of the feedback provided by the secondary recirculating bubble. I'_4 represents the largest destabilizing factor in the integral energy budget which describes an amplification process due to basic-flow deceleration (cf. table 4.8). However, its importance is qualified because I'_2 is strongly stabilizing in the same region where I'_4 features its maximum production (figures 4.31c,d). By integrating over the absolute value of the lo-

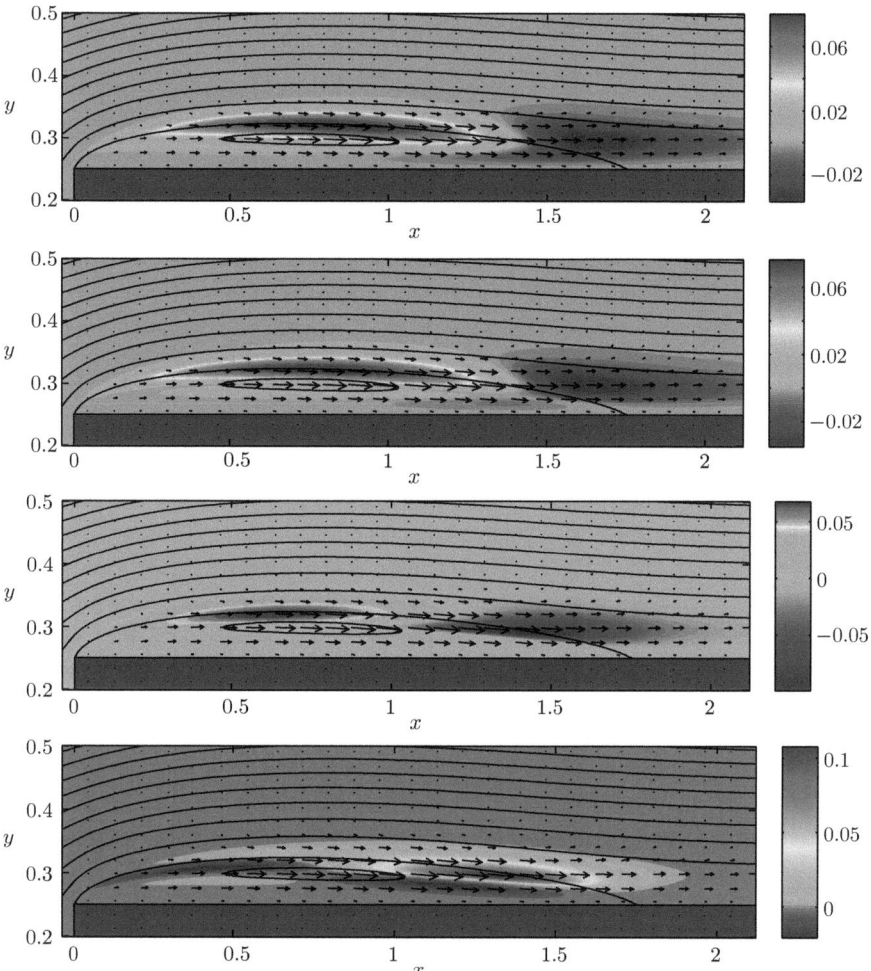

Figure 4.31.: Critical velocity fields (arrows) and local energy-production rates for $\Gamma^{\mathrm{f}} = 0.25$ in a plane $z = 0$ in which all production terms shown exhibit their local extrema: (a) total local production $\sum_i I_i$, (b) $I'_2 + I'_4$, (c) I'_2 and (d) I'_4. The y axes are stretched by a factor of two.

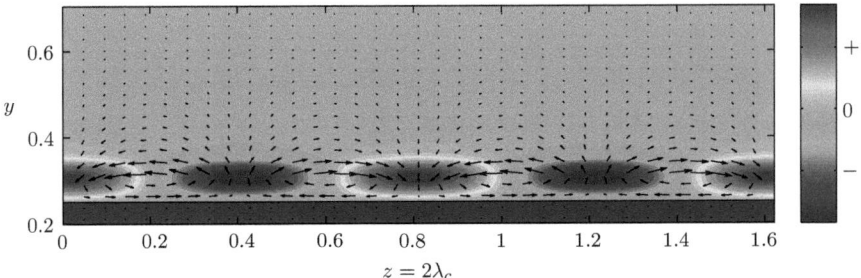

Figure 4.32.: Critical velocity fields (arrows) and streamwise perturbation velocity $\tilde{u}$ (colour) for $\Gamma^{\mathrm{f}} = 0.25$ at $x = 0.8342$, where the energy production features its local maximum.

cal energy-production rates $\int_V |I_i'| \mathrm{d}V$, one can see that the globally most important effect is represented by I_2' (last row of table 4.8). The term I_2' describes the classical lift-up effect by which basic-flow momentum is transported by the cross-stream perturbation velocity $\tilde{\boldsymbol{u}}_\perp \cdot \nabla \boldsymbol{u}_0$, feeding back on the parallel component $\tilde{\boldsymbol{u}}_\parallel$. By comparing figure 4.31a with figure 4.31b, one can hardly notice any difference between the sum of all energy-production rates $\sum I_i'$ and the sum of $I_2' + I_4'$.

To sum up, the physical nature of the instability is a combination of the lift-up mechanism and flow deceleration near the separating streamline of the secondary recirculating vortex, providing a feedback for the disturbances to be self-sustained.

Sensitivity results

A sensitivity analysis was also performed for $\Gamma^{\mathrm{f}} = 0.25$ at critical conditions. From figure 4.35, it can be seen that the flow is most sensitive to initial conditions and momentum forcing just at the entrance of the channel extending forward to the step front. A certain receptivity is also found after the step within the secondary separation zone.

The product between the direct and the adjoint fields $\Upsilon = |\hat{\boldsymbol{u}}| \, |\hat{\boldsymbol{u}}|$ (2.4.8) is evaluated in figure 4.36, showing the regions where the leading eigenvalue and, consequently, the stability boundary is most sensitive to local perturbations. It can be seen that the highest effect on the growth rate and thus the stability boundary is provided by perturbations localized in the region of the secondary recirculating vortex. This result is consistent with the above energy analysis, which has also shown that the instability is caused by quite localized processes after the step and within the region of the secondary recirculating bubble.

Stüer et al. (1999) observed by hydrogen-bubble visualisation that the turbulent regime for $\Gamma^{\mathrm{f}} = 0.25$ sets in at around $\mathrm{Re} \approx 4800$. This difference of about $-20\,\%$ with respect to the global stability boundary may be related to sidewall effects and/or to the sensitivity of the flow to oncoming perturbations, which both cannot be avoided in laboratory experiments.

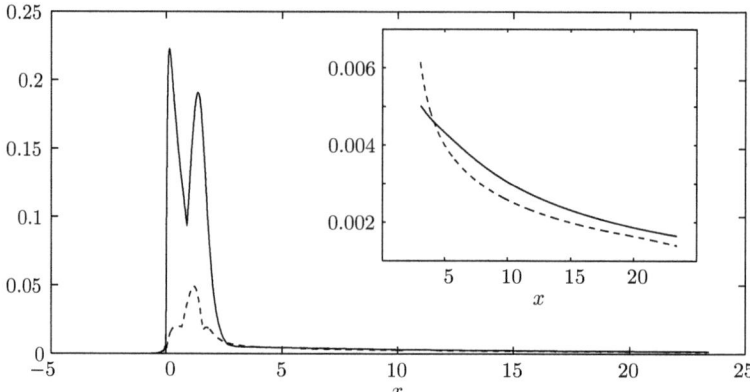

Figure 4.33.: Relative magnitude of the streamwise vortices expressed by $\max_y |\hat{v}(x,y)|$ (dashed) and $\max_y |\hat{w}(x,y)|$ as functions of the streamwise coordinate x for $\Gamma^f = 0.25$ at $\mathrm{Re}_c = 5888.4$. The inset shows the decay of the amplitudes for $x \in [3; 24]$.

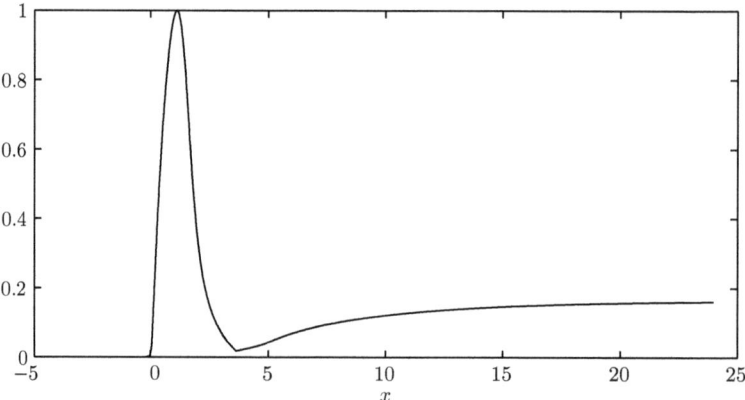

Figure 4.34.: Amplitudes of the streamwise streaks expressed by $\max_y |\hat{u}(x,y)|$ as functions of the streamwise coordinate x for $\Gamma^f = 0.25$ at $\mathrm{Re}_c = 5888.4$.

4.2.4. Conclusion

The global linear stability of the flow past a forward-facing step has been carried out for a quasi-continuous variation of the constriction ratio Γ^f. The influence of the grid resolution and of the in- and outlet-channel lengths on the stability boundaries has been analysed to ensure independent results. The critical Reynolds number Re_c takes its maximum value for the smallest constriction ratio considered. Re_c decreases gradually as the step height is increased. By using a proper length scale, the scaled critical values Re_c^* and k_c^* approach a linear asymptotic regime for large step heights $(1 - \Gamma^f) \ll 1$. All three-dimensional critical modes are stationary and locally confined to the secondary recirculation bubble, which is thus identified as the main source of the instability. An *a posteriori* energy-transfer analysis showed that all instabilities can be explained in terms of the combined action of lift-up and flow deceleration.

The critical mode and its associated energy-production rates were visualized for $\Gamma^f = 0.25$. The critical wavelength is approximately three times the step height, in agreement with the experimental and numerical results of Stüer *et al.* (1999) and Wilhelm *et al.* (2003). Almost all of the kinetic perturbation energy is produced near the separating streamline of the secondary vortex of the basic flow. The critical mode appears as steady, streaky structures downstream of the step. This streak formation has also been observed in the experiments of Pollard *et al.* (1996) and Stüer *et al.* (1999). The alternating slow and fast streaks of the perturbations are self-sustained here due to the feedback provided by the recirculating bubble. The sensitivity analysis for spatially localized perturbations likewise identified the secondary recirculating bubble to be extremely important for the temporal growth rate and thus for the stability boundary. It is worth mentioning that the topology of the critical mode is very similar to the one detected in the backward-facing-step problem for small step heights.

A problem for the experimental observability of the global instability of the forward-facing-step problem is the sensitivity of the flow with respect to oncoming disturbances, which was revealed by the absolute value of the adjoint of the critical mode. An open question is the

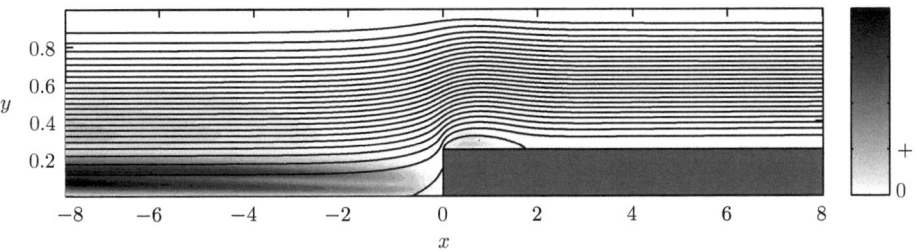

Figure 4.35.: Receptivity to initial conditions and momentum forcing $|\hat{u}|$ for the most dangerous mode for $\Gamma^f = 0.25$ at $Re_c = 5888.4$.

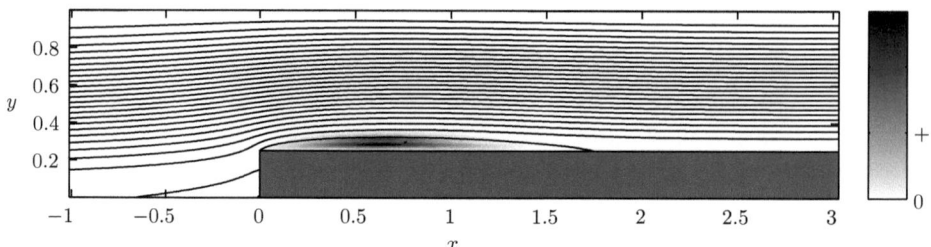

Figure 4.36.: Receptivity Υ to spatially localized feedback (2.4.8) for the most dangerous mode for $\Gamma^{\mathrm{f}} = 0.25$ at $\mathrm{Re}_c = 5888.4$.

existence of convective instabilities below the absolute stability boundaries as presented in the current work. Transient-growth analyses in a global framework to be conducted in future studies may contribute to answer this question. Also, direct numerical simulations and experiments focusing on the global three-dimensionality would be of interest to assess the physical relevance of the global instability modes.

4.3. The Plane Sudden-Expansion Problem

4.3.1. Problem Formulation

An incompressible, Newtonian fluid is considered in a plane channel with a sudden expansion. The system is considered to be infinitely extended in the spanwise (z) direction. The geometry is sketched in figure 4.37, consisting of an inlet channel of length L_i and height h_i, followed by a suddenly expanded channel of height H and length L_o. The lower step height is denoted by h_l and the upper one by h_u. The origin of the Cartesian coordinate system is located at the bottom of the outlet channel and at the sudden expansion.

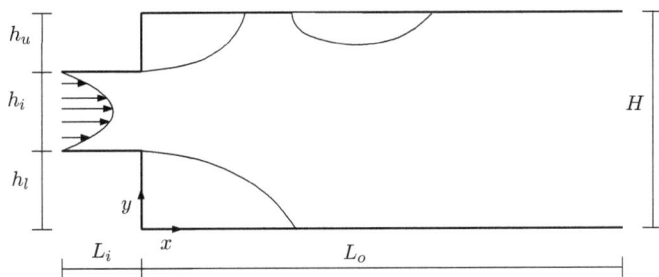

Figure 4.37.: Sketch of the channel geometry with a asymmetric sudden expansion and separating streamlines.

The geometry is characterized by the expansion ratio

$$\Gamma^e = \frac{h_l + h_u}{H} = 1 - \frac{h_i}{H} \tag{4.3.1}$$

and the asymmetry ratio

$$\alpha = \frac{|h_l - h_u|}{h_s}, \tag{4.3.2}$$

where $h_s = (h_l + h_u)/2$ denotes the average step height (symmetric case).

To conform with previous investigations and for computational economy, it is useful to define the Reynolds number as

$$\mathrm{Re} = \frac{LU_\infty}{\nu}. \tag{4.3.3}$$

It is based on half of the *downstream* channel height $L = H/2$ and the maximum (centreline) velocity U_∞ of the *inlet* profile. With this definition of Re, the generation of the various grids is convenient because the outlet channel can be kept constant and only the inlet channel must be varied.

No-slip and no-penetration boundary conditions $\boldsymbol{u}_0 = 0$ are imposed at the channel walls. At the inlet $x = -L_i/L$, a fully developed laminar plane Poiseuille flow $u_0(y) = 4(y - h_l)(h_i - y + h_l)/h_i^2$ and $v_0 = 0$ is assumed for $y = h_i$. At the outflow boundary $x = L_o/L$, the normal derivatives of all velocity components are set to zero, i.e. $\partial_x \boldsymbol{u}_0 = 0$, and the pressure is specified as $p_0 = 0$.

Assuming that the inlet L_i as well as the outlet channel lengths L_o are sufficiently large such that the results are independent of L_i and L_o, the problem is governed by three parameters: the Reynolds number Re (4.3.3), the expansion ratio Γ^e (4.3.1) and the asymmetry α (4.3.2) of the geometry. The dependence of the stability boundaries on the channel lengths will be considered in more detail in §4.3.3.

4.3.2. Scientific Background

The steady laminar two-dimensional flow of an incompressible Newtonian fluid though a channel with a symmetric sudden expansion can either be symmetric or asymmetric with respect to the plane of symmetry, depending on the Reynolds number. Experiments of Durst et al. (1974), Cherdron et al. (1978), Sobey & Drazin (1986), Fearn et al. (1990) and Durst et al. (1993) have shown that the primary flow is steady, two-dimensional and symmetric, with two recirculation zones of equal size near the expansion corners if the expansion ratio is moderate and the Reynolds number Re is sufficiently small. The length of the separation zones increases linearly with the Reynolds number. At higher Re, however, the flow loses its symmetry with respect to the mid-plane and a pair of stable, steady, two-dimensional asymmetric flow states evolves gradually with recirculation zones of various lengths.

Using experimental and numerical techniques, Fearn et al. (1990) demonstrated that the symmetric solution (primary flow) becomes unstable at a primary critical Reynolds number, and a pair of stable, asymmetric solutions (secondary flow) bifurcates supercritically. The slightly imperfect bifurcation found experimentally by Fearn et al. (1990) was modelled numerically by introducing a slightly asymmetric expansion in the computational geometry.

The linear stability analyses of Shapira & Degani (1990) and Alleborn et al. (1997) for a symmetric channel geometry revealed that the symmetric primary flow is stable for $\text{Re} \leq \text{Re}_c$ with respect to two-dimensional perturbations. At higher Reynolds numbers, however, it loses its stability via a symmetric pitchfork bifurcation, and two stable, asymmetric secondary flow states evolve. This result was confirmed by Rusak & Hawa (1999) by use of a weakly nonlinear analysis of the unsteady Navier–Stokes equations for $|\text{Re} - \text{Re}_c|/\text{Re} \ll 1$. The two-dimensional, time-dependent simulations of Hawa & Rusak (2001) established the relationship between the neutral modes of the linear stability analysis and the nonlinear time-asymptotic secondary flow. The asymptotic analysis of Rusak & Hawa (1999) was extended by Hawa & Rusak (2000) by considering a slightly asymmetric expansion. Above a certain threshold Reynolds number, three asymmetric solutions were found to exist: a stable solution branch continuously evolving out of the primary flow and two disconnected solutions, one stable and the other unstable, which are created through a saddle-node bifurcation. The contiguous smooth solution branch of the imperfect pitchfork bifurcation showed close agreement with the experimental data of Fearn et al. (1990), whereas the stable saddle-node branch did not agree well with the corresponding unperturbed secondary flow. These results were confirmed by Mizushima & Shiotani (2000) in their weakly nonlinear stability analysis for a slightly imperfect channel geometry.

Three-dimensional, time-dependent simulations were performed by Schreck & Schäfer (2000), Chiang et al. (2000), Chiang et al. (2001) and Tsui & Wang (2008), taking into account the finite spanwise extent of the channel. Schreck & Schäfer (2000) reported that narrower channel depths, i.e. smaller spanwise aspect ratios, decrease the lengths of the recirculation zones and stabilise the primary symmetric and nominally two-dimensional flow, which was also observed experimentally by Cherdron et al. (1978). Chiang et al. (2000) also studied the sidewall effects on the structure of the laminar flow for various spanwise aspect ratios $\Lambda = d/h_i$, i.e. depth-to-height of the inflow channel. They reported that the two-dimensional results differ considerably from the three-dimensional ones for $\Lambda \leq 12$. Recently, Tsui & Wang (2008) confirmed the findings of Chiang et al. (2000), and noted that for $\Lambda \geq 24$ the flow structure can be regarded as two-dimensional as the sidewall effects are negligible in the central part of the channel. In addition to the symmetry breaking of the two-dimensional flow, the simulations of Chiang et al. (2001) revealed the existence of a spanwise modulated three-dimensional flow for certain initial conditions. The flow with spanwise modulation was found to be symmetric with respect to the mid-span plane as well as with respect to the mid-channel plane. Since the corresponding two-dimensional solution, which is symmetric with respect to the mid-channel plane, is unstable, the modulated three-dimensional flow is difficult to obtain numerically and could only be observed for $\Lambda \geq 30$.

To date, all linear stability analyses have dealt with the first bifurcation of the primary flow to the stationary secondary flow. No results are available on the stability of the asymmetric two-dimensional secondary flows. Fearn et al. (1990) and Battaglia et al. (1997) anticipated that the asymmetric solutions would become unstable to three-dimensional perturbations. An alternative was suggested by Cherdron et al. (1978) and Durst et al. (1993), in which a Hopf bifurcation may result in a time-periodic two-dimensional state. To answer this open question, a three-dimensional global linear stability analysis is carried out. In doing so the geometry is varied in a quasi-continuous way, covering expansion ratios (step-to-outlet height ratios) from 0.25 to 0.95. It is shown that, within this parameter range, the primary instability of the symmetric basic flow is two-dimensional. Moreover, the asymmetric two-dimensional solutions become unstable to three-dimensional and not to two-dimensional perturbations. In case of an asymmetric geometry the disconnected saddle-node-bifurcation point for the disconnected steady two-dimensional solution branches is shifted to higher Reynolds numbers as the channel

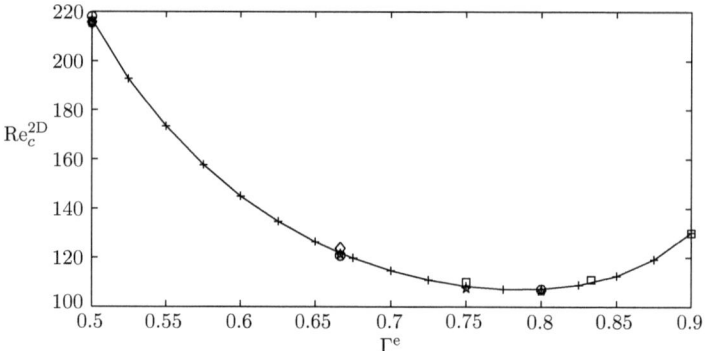

Figure 4.38.: Critical Reynolds number for the two-dimensional instability of the symmetric flow for $\alpha = 0$ (line and plus signs) as function of the expansion ratio with $L_o = 25L$. Also shown are the results of Shapira & Degani (1990) (diamonds), Alleborn et al. (1997) (circles), Battaglia et al. (1997) (stars) and Drikakis (1997) (squares).

asymmetry is increased (see figure 4.40). Hence, even for a slight asymmetry, only the non-symmetric solution of the contiguous branch is relevant for the three-dimensional stability analysis. Its stability boundaries are found to be similar to those of the asymmetric solutions for the symmetric channel flow.

4.3.3. Results

Grid independence and code verification

The results were found to be grid-independent if the grid resolution is at least $\Delta x_{\min} = \Delta y_{\min} = 0.005$ (in units of L) along the walls and around the step. The cell width is increased smoothly with a constant stretching factor of 1.03 until the maximum grid spacings $\Delta x_{\max} = 0.075$ or $\Delta y_{\max} = 0.025$ are reached. This meshing strategy maintains a relatively constant cell density across the inflow channel. Relevant grid parameters for representative expansion ratios can be found in table 4.10.

For verifying the numerical code, the two-dimensional stability boundaries for the symmetric solution are compared with those obtained by Shapira & Degani (1990), Alleborn et al. (1997), Battaglia et al. (1997) and Drikakis (1997). The result is depicted in figure 4.38. The data, computed with an outflow-channel length of $L_o = 25L$, are consistent and show a very good agreement.

Dependence on channel lengths

The influence of the entrance length L_i and the length of the outflow channel L_o on the stability boundaries was analysed in order to ensure that the critical data are independent of these two geometry parameters. Different critical Reynolds numbers arise. For clarity, the critical Reynolds numbers (as well as wave numbers) for the two-dimensional instability of the symmetric two-dimensional basic flow in the symmetric geometry (primary instability for $\alpha = 0$) will be denoted by Re_c^{2D}, the three-dimensional critical Reynolds number of the asymmetric two-dimensional basic flow in the symmetric geometry (secondary instability for $\alpha = 0$) is denoted by Re_c^0, and the critical Reynolds number for the continuous two-dimensional basic flow in the perturbed (asymmetric, $\alpha \neq 0$) geometry is called Re_c^α.

The effect of the inlet-channel length L_i on the stability boundaries is shown in table 4.9. Results are reported for $\Gamma^e = 0.5$ for the asymmetric solution of the symmetric channel Re_c^0, but a similar convergence of the critical data, as L_i is increased, holds true for all expansion ratios covered here, and also for the asymmetric channel geometry. As can be observed from table 4.9, the entrance length should be at least $L_i = 6h_i$. In the current study the inlet-channel length was set to $L_i = 8h_i$ to keep its effect on the critical data below $\approx 0.02\%$. Such a long entrance length is necessary because the critical mode extends well upstream of the expansion, which is shown in figure 4.39.

The influence of the outflow length L_o on the results was also investigated. The required outflow channel length L_o depends on the Reynolds number (see table 4.10). It must be long enough for a plane Poiseuille flow to be developed at the outlet. The outlet length is selected such that the maximum relative deviation of the x-component of the outlet velocity profile from a plane Poiseuille flow is less than 2 %. Note that if L_o is too short, unphysical modes may arise for high Krylov-subspace dimensions (≥ 150), being artefacts of the outlet boundary conditions.

Stability boundaries

Since the basic two-dimensional flow beyond the primary instability threshold Re_c^{2D} is not unique, care must be taken to find the symmetric and asymmetric basic flows by Newton iteration. To stay on the basic symmetric-flow solution branch for $\alpha = 0$ and $\mathrm{Re} > \mathrm{Re}_c^{2D}$ it suffices to increase the Reynolds number stepwise by less than 10 % using the previously

L_i	$1h_i$	$2h_i$	$3h_i$	$4h_i$	$6h_i$	$8h_i$	$15h_i$
Re_c^0	806.47	797.17	793.25	791.63	790.98	790.94	790.94
k_c^0	0.6615	0.6490	0.6387	0.6340	0.6318	0.6317	0.6317

Table 4.9.: Critical values Re_c^0 and k_c^0 as functions of the entrance length L_i for multiples of the inlet-channel height h_i for $\Gamma^e = 0.5$ with $L_o = 65$.

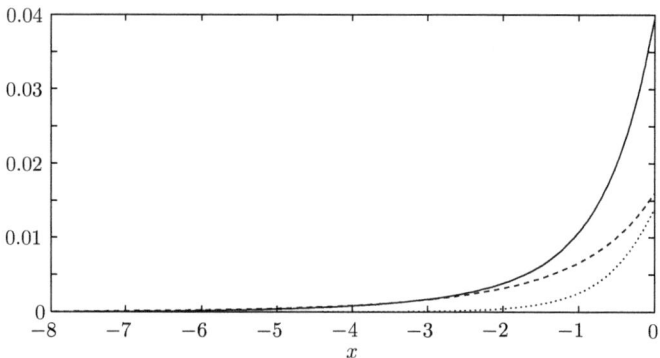

Figure 4.39.: Maxima of the velocity-perturbation components in the inlet channel for $\Gamma^e = 0.5$ at $\mathrm{Re}_c^0 = 790.94$ and $k_c^0 = 0.6317$. The solid, dotted and dashed lines represent $\max_y |\hat{u}(x,y)|$, $\max_y |\hat{v}(x,y)|$ and $\max_y |\hat{w}(x,y)|$, respectively, in units of (2.2.6).

Γ	Re_c^{2D}	L_o	N_x^o	N_x^i	N_y^i	N_y^o
0.25	1657.7	126	1771	223	116	192
0.3	934.83	78	1119	212	112	200
0.4	394.87	34	522	190	104	212
0.5	216.76	25	399	169	94	218
0.6	144.96	20	331	147	82	222
0.6̇	121.74	20	331	132	74	222
0.7	114.59	10	196	125	70	222
0.8	107.11	10	196	103	54	218
0.9	130.17	10	196	80	32	208
0.95	188.47	10	196	59	18	202

Γ	Re_c^0	k_c^0	$|\omega_c^0|$	Re_c^α	k_c^α	$|\omega_c^\alpha|$	L_o	N_x^o	N_x^i	N_y^i	N_y^o
0.25	6434.2	0.1292	0	6319.5	0.1294	0	400	5491	223	116	192
0.3	3789.5	0.1421	0	3728.4	0.1427	0	320	4405	212	112	200
0.4	1569.5	0.4766	0	1542.1	0.5301	0	130	1825	190	104	212
0.5	790.95	0.6339	0	774.43	0.6715	0	65	943	169	94	218
0.6	513.31	0.8064	0	503.91	0.8249	0	40	603	147	82	222
0.6̇	454.83	0.9683	0	447.81	0.9713	0	40	603	132	74	222
0.7	445.80	1.1726	0	441.13	1.1571	0	35	535	125	70	222
0.8	375.98	1.3262	0.0253	374.16	1.3413	0.0259	25	399	103	54	218
0.9	370.53	1.3066	0.0456	373.65	1.3492	0.0470	15	264	80	32	208
0.95	423.48	1.3031	0.0485	430.85	1.3582	0.0509	15	264	59	18	202

Table 4.10.: Critical parameters for the scaling (4.3.3) for selected expansion ratios Γ^e, the outflow length L_o and the number of grid points (N_x, N_y) in the (x, y) plane.

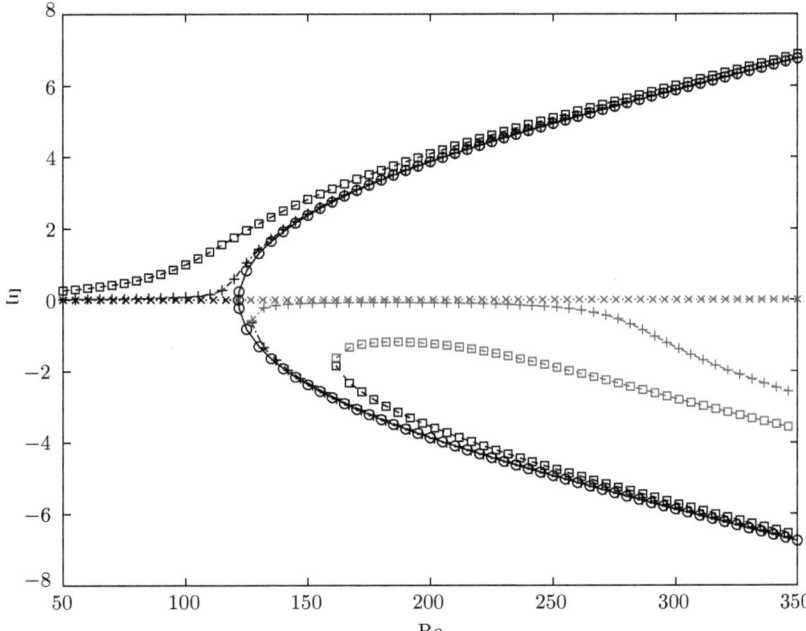

Figure 4.40.: The measure of asymmetry (4.3.4) for various basic-flow solutions. Crosses and dots represent the symmetric flow, and cricles and solid lines indicate the asymmetric flow solutions for the symmetric channel $\alpha = 0$. Plus signs and dash-dotted lines denote the flow solutions of the slightly asymmetric geometry $\alpha = 0.01$, whereas squares and dashed lines represent those for $\alpha = 0.15$.

computed solution as the initial guess. To find and track one of the two asymmetric solutions $u_0 \equiv v_0 \equiv 1$ and $p_0 \equiv 0$ is utilized as the initial guess for the Newton iteration, and the Reynolds number is selected as $\text{Re} > 1.2 \times \text{Re}_c^{2D}$, using the same step size in Re as for the symmetric-solution tracking.

Figure 4.40 shows for $\Gamma^e = 0.\dot{6}$ the topological change of the pitchfork-bifurcation diagram upon an asymmetric geometry perturbation as described in §4.3.2. The following measure of asymmetry is chosen to be

$$\Xi = \text{sgn}(x_l - x_u)\sqrt{\int (\partial_y u_0|_{y=H} + \partial_y u_0|_{y=0})^2 dx}, \quad (4.3.4)$$

where x_l and x_u denote the lengths of the lower and the upper primary recirculation bubbles, respectively, and the wall shear stresses are evaluated at the upper and lower walls of the outlet channel. For a symmetric geometry ($\alpha = 0$) the basic symmetric-flow solution, depicted by crosses and dots, becomes unstable at $\text{Re}_c^{2D} = 121.74$. At this critical Reynolds number two stable and asymmetric flow solutions (circles and solid lines) bifurcate via a perfect pitchfork bifurcation. If the expansion is slightly asymmetric ($\alpha \neq 0$), the bifurcation becomes imperfect (plus signs and dash-dotted lines) and a saddle-node bifurcation arises at $\text{Re}^{\text{SN}}(\alpha = 0.01) \approx 127$. The disconnected saddle-node-bifurcation point is shifted to higher Reynolds numbers as the channel asymmetry is increased. This is shown for $\alpha = 0.15$ by squares and dashed lines, where the saddle-node-bifurcation point is located at $\text{Re}^{\text{SN}}(\alpha = 0.15) \approx 161.5$. In this case, the connected solution branch deviates substantially from the flow solutions corresponding to $\alpha = 0$ in the vicinity of $\text{Re}_c^{2D} = 121.74$. However, the differences are getting smaller as the Reynolds number is increased. The upper solution branch emerging from the saddle-node-bifurcation point is unstable from the beginning, which is indicated by red colour. The substantial deviation of Ξ from zero for the unstable supercritical solution for $\alpha = 0.01$ at Reynolds numbers $\text{Re} \approx > 250$ is due to the creation of a third recirculation bubble.

In the present study, the attention is restricted to the connected solution branch with $\alpha = 0.15$. In fact, this solution for $\alpha = 0.15$ is very similar to the corresponding asymmetric solution for $\alpha = 0$, i.e. for the symmetric geometry (see figures 4.41 and 4.42).

Representative numerical critical Reynolds numbers, wave numbers and oscillation frequencies can be found in table 4.10. Data are provided for selected expansion ratios for the primary instability (Re_c^{2D}) and the secondary instability (Re_c^0), both in the symmetric channel, and also for the instability of the asymmetric flow of the connected solution branch (Re_c^α) in the asymmetric geometry ($\alpha = 0.15$). Moreover, the outflow lengths L_o and the number of grid points are specified.

For an overview of the critical data, they are rescaled in the prevailing subsection 4.3.3 using the inlet velocity U_∞ and inlet length scale $h_i/2$, yielding

$$\text{Re}^* = \text{Re}\frac{h_i}{H}, \qquad k^* = k\frac{h_i}{H}, \qquad \omega^* = \omega\frac{h_i}{H}. \qquad (4.3.5)$$

The inlet Reynolds number $\text{Re}^* = U_\infty^{\text{inlet}} h_i/(2\nu) = U_\infty^{\text{outlet}} H/(2\nu)$ is equivalent to the one at the outlet.

Figure 4.41 shows various critical inlet Reynolds numbers (Re_c^{2D*}, Re_c^{0*} and $\text{Re}_c^{\alpha*}$, respectively) on a logarithmic scale as functions of the expansion ratio for $\Gamma^e \leq 0.5$. In the range shown, all critical modes are stationary and the stability boundaries scale nearly exponentially with Γ^e. The stability boundaries of the asymmetric flow in the symmetric channel Re_c^{0*} are very similar to those of the solution in the asymmetric channel $\text{Re}_c^{\alpha*}$ for $\alpha = 0.15$, which can be seen by the two overlapping curves in the figures 4.41 and 4.42. Note that the gap between the curves at $\Gamma^e \in [0.337; 0.35]$ should indicate that the critical modes and their underlying instability mechanisms change (see also figure 4.43). Moreover, the critical Reynolds number for the symmetric flow Re_c^{2D*} is smaller than Re_c^{0*} and $\text{Re}_c^{\alpha*}$ by approximately a factor of $1/4$.

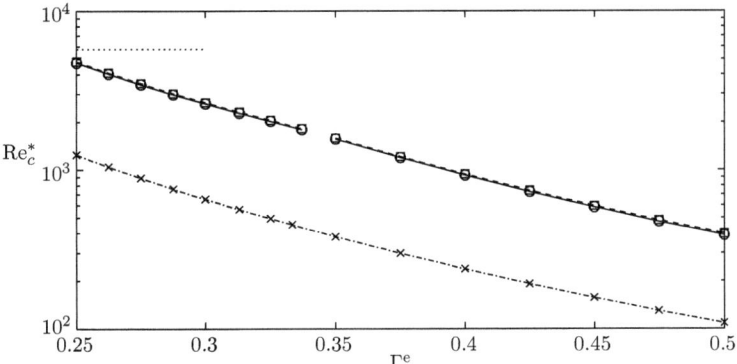

Figure 4.41.: Scaled critical inlet Reynolds numbers Re_c^* as functions of the expansion ratio Γ^e. Crosses and the dash-dotted line represent Re_c^{2D*}, open circles and the full line indicate Re_c^{0*}, and open squares and the dashed line denote $\mathrm{Re}_c^{\alpha*}$ for $\alpha = 0.15$. The dotted line indicates the critical Reynolds number of plane Poiseuille flow. All Reynolds numbers are shown on a logarithmic scale.

Even for the smallest step height considered, i.e. $\Gamma^e = 0.25$, all critical Reynolds numbers are below the linear stability boundary $\mathrm{Re}_c^{\mathrm{Poiseuille}} = 5772$ of plane Poiseuille flow, which is indicated by a dotted line in figure 4.41.

As the step height is increased, the critical inlet Reynolds number decreases continuously and scales linearly for very large expansion ratios $(1 - \Gamma^e) \ll 1$, as shown in figure 4.42. The linear asymptotic behaviour for $(1 - \Gamma^e) \ll 1$ is estimated as $\mathrm{Re}_c^{2D*} \approx 77.60 - 71.75 \times \Gamma^e$ and $\mathrm{Re}_c^{0*} \approx \mathrm{Re}_c^{\alpha*} \approx 331.64 - 326.43 \times \Gamma^e$, respectively. As can be seen from figure 4.42, the difference in the critical Reynolds numbers between the symmetric (Re_c^{2D*}) and asymmetric solutions (Re_c^{0*} and $\mathrm{Re}_c^{\alpha*}$) decreases continuously (also relatively) as Γ^e is increased.

It is found that the critical modes corresponding to Re_c^{0*} and $\mathrm{Re}_c^{\alpha*}$ change twice in the range of expansion ratios considered. Since the slopes of the intersecting neutral curves are nearly the same, the modal change is more easily seen in the jump of the critical wave numbers which are shown in figure 4.43. The first modal change is among stationary modes and it arises in the interval $\Gamma^e \in [0.337; 0.35]$. The second modal change is from a steady to an oscillatory mode and occurs within $\Gamma^e \in [0.7; 0.713]$. From figure 4.43 it is also observable that the two wave numbers k_c^{0*} and $k_c^{\alpha*}$ remain approximately constant (≈ 0.1) for small step heights and scale linearly for large expansion ratios, which is estimated as $k_c^{0*} \approx k_c^{\alpha*} \approx 1.33 - 1.32 \times \Gamma^e$. The scaled frequencies ω_c^* of the critical modes are shown in figure 4.44 for large expansion ratios $\Gamma^e > 0.7$. The asymptotes are estimated by $\omega_c^{0*} \approx 0.05 \times (1 - \Gamma^e)$ (red dots) and $\omega_c^{\alpha*} \approx 0.051 \times (1 - \Gamma^e)$ (blue dots), respectively.

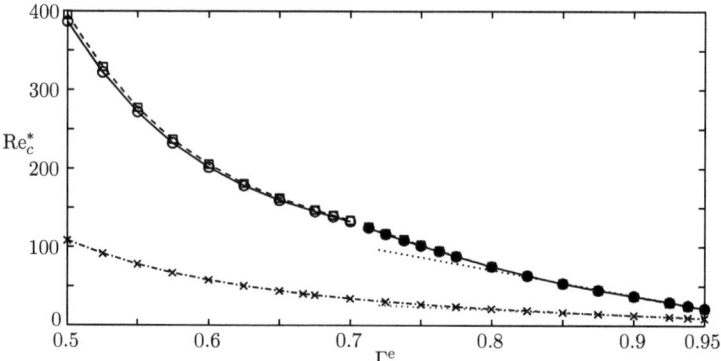

Figure 4.42.: Scaled critical inlet Reynolds numbers Re_c^* as functions of the expansion ratio Γ^e. Crosses and the dash-dotted line represent Re_c^{2D*}, circles and the full line indicate Re_c^{0*}, and squares and the dashed line denote $\mathrm{Re}_c^{\alpha*}$ for $\alpha = 0.15$. Open symbols stand for stationary modes, full ones for oscillatory modes. The dotted lines indicate the linear behaviour for $\Gamma^e \to 1$ as mentioned in the text.

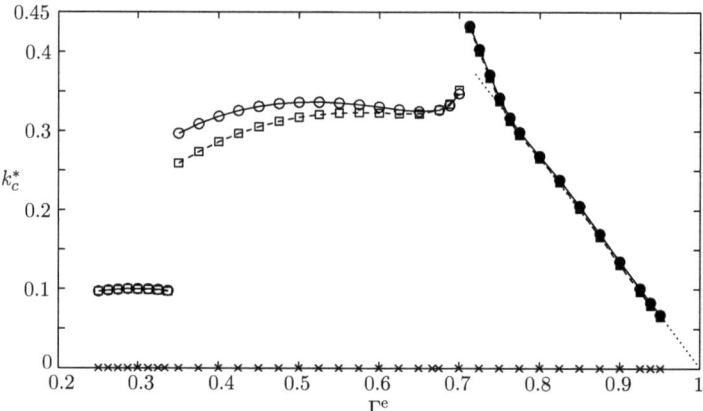

Figure 4.43.: Scaled critical wave numbers k_c^* as functions of the expansion ratio Γ^e. Crosses represent k_c^{2D*}, circles and the full line k_c^{0*} and squares and the dashed line $k_c^{\alpha*}$ for $\alpha = 0.15$. Open symbols stand for stationary modes, full ones for oscillatory modes. The dotted line indicates the linear behaviour as mentioned in the text.

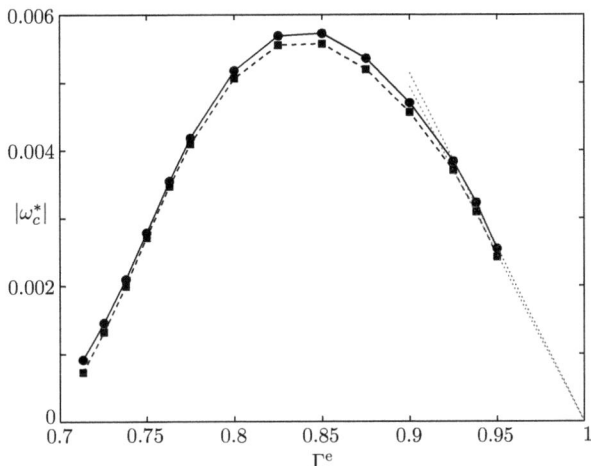

Figure 4.44.: Scaled critical frequencies ω_c^* as functions of the expansion ratio Γ^e. Circles and the full line represent ω_c^{0*}, squares and the dashed line $\omega_c^{\alpha*}$ for $\alpha = 0.15$. The dotted lines indicate the asymptotes for $\Gamma^e \to 1$ as mentioned in the text.

Regardless of the expansion ratio Γ^e, the two-dimensional basic flow immediately behind the expansion is characterized by two separated recirculation zones which are symmetric for $\alpha = 0$ and $\text{Re} \leq \text{Re}_c^{2D}$. They are referred to as primary vortices. For the asymmetric basic flows at higher Reynolds numbers three recirculation bubbles are present. Two of them separate immediately downstream of the sharp corners. The stronger and larger of theses vortices will be called primary, the weaker and smaller vortex on the opposite side of the channel the secondary vortex. On the same side of the weaker vortex further downstream, a third separated vortex arises. Even though the computational domain is much more extended in the streamwise (x) direction, the flow characteristics shall be discussed only near the separated vortices, because these regions are found to be of key importance for the flow instabilities. It is these regions where the critical modes and the associated energy-transfer rates are most pronounced.

Symmetry-breaking bifurcation in the symmetric expansion

The physical nature of the two-dimensional instability of the primary, symmetric flow for $\alpha = 0$ is nearly independent of the expansion ratio. As a representative case for a strong expansion, the primary, symmetry-breaking instability for $\Gamma^e = 0.9$ is considered. The critical Reynolds number is $\text{Re}_c^{2D} = 130.17$. As can be seen from figure 4.45, the basic flow is symmetric and represents a jet into a laterally bounded domain. The counter-rotating recirculation zones are approximately three times as long as the step height. The critical mode appears as a single vortex centred at the axis of symmetry $y = 1$ and spanning the full width of the outlet channel.

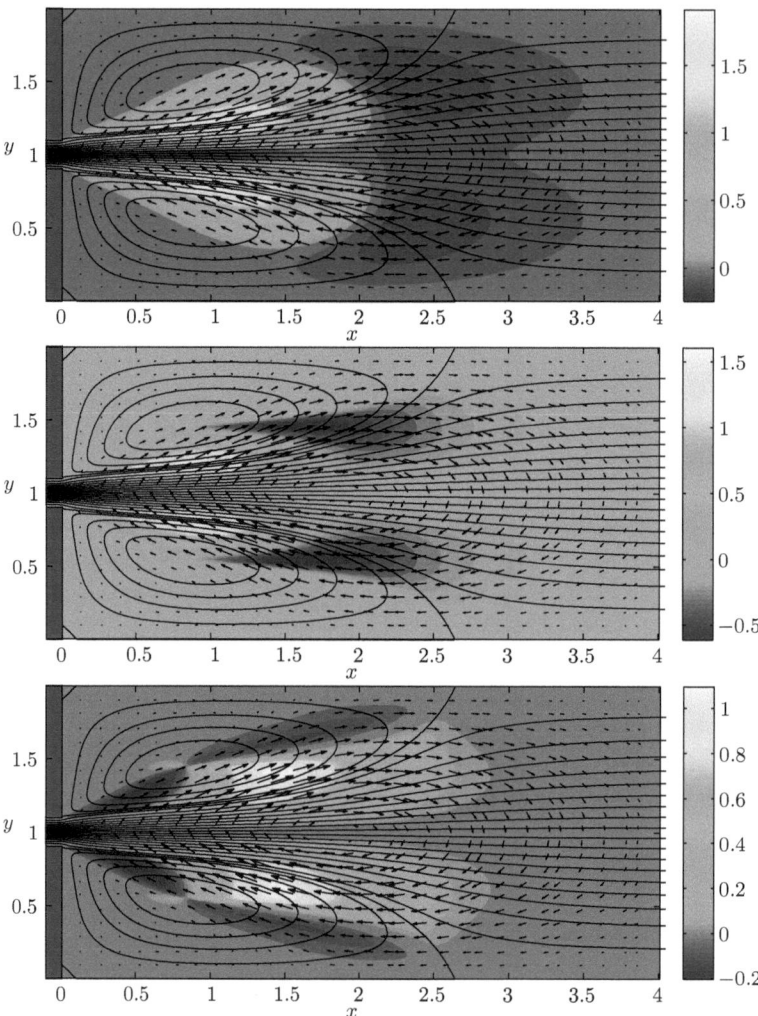

Figure 4.45.: Basic-state streamlines, critical velocity fields (arrows) and local energy production rates for $\Gamma^e = 0.9$ and $\mathrm{Re}_c^{2D} = 130.17$: (a) total local production $\sum_i I'_i$, (b) I'_2 and (c) I'_4.

i	1	2	3	4	Γ^e
$\int_{\mathcal{D}} I'_i d\mathcal{D}$	−0.1480	0.3577	0.0316	0.7587	0.9
$\int_{\mathcal{D}} I'^{+}_i d\mathcal{D}$	0.0355	0.4322	0.0625	0.4699	0.9
$\int_{\mathcal{D}} I'_i d\mathcal{D}$	−0.0474	−0.1418	0.0441	1.1451	0.25
$\int_{\mathcal{D}} I'^{+}_i d\mathcal{D}$	0.0149	0.8057	0.0067	0.1727	0.25

Table 4.11.: Global normalized energy production rates for the symmetry-breaking bifurcation. $\int_{S_o} I_5 dS = 0$ for all cases considered.

From figure 4.45a it can be noticed that most of the kinetic energy is transferred from the basic state to the perturbations in the shear-layers near the separating streamlines. The total local energy transfer peaks at around $x \approx 1$.

The largest integral contribution to the energy transfer is due to I'_4 (see table 4.11), which describes the streamwise transport of basic flow momentum $\tilde{u}_\parallel \cdot \nabla u_0$, feeding back on the streamwise perturbation flow $\tilde{u}_\parallel$. In order for this process to have a destabilizing effect, the basic flow must decelerate, i.e. $e_\parallel \cdot \nabla u_0 < 0$. The term I'_4 is most pronounced within the separation zones with maximum energy production near $x \approx 1.5$. Its importance, however, is quantified, because the process I'_2 is locally stabilising in the same region (see figure 4.45b, c). The term I'_2 exhibits the strongest local maxima, which are reflected in the total local production (figure 4.45a). The amplification process I'_2 requires a high shear rate, which is present in the shear-layers between the separation zones and the bulk flow.

Table 4.11 shows the global normalized energy production rates for the symmetry-breaking bifurcation. The integrals are also evaluated over the flow domain $\mathcal{D}$ restricted to positive values of the integrands I'^{+}_i. This yields their relative contribution to the destabilising energy transfer. It can be noticed that the terms I'^{+}_2 and I'^{+}_4 are indeed of comparable importance. Thus the two-dimensional symmetry-breaking instability at $\Gamma^e = 0.9$ is based on the combined effects of flow deceleration and shear-layer instability.

It is worth mentioning that Shtern & Hussain (2003) found similar instabilities for conical jets. They showed via a non-parallel analysis that flow deceleration plays a major role for swirl-free jets, triggering non-axisymmetric instabilities.

When the step height is decreased, the primary vortices become very much elongated downstream. For $\Gamma^e = 0.25$, for instance, the recirculating zones are approximately 34 times as long as the step height and the bulk flow is almost parallel (see figure 4.46). Again, the critical mode arises as a single stationary vortex centred at the line of symmetry, but this vortex is much more elongated than in the case of $\Gamma^e = 0.9$.

As can be seen from table 4.11, the term I'^{+}_2 is clearly dominating and I'^{+}_4 plays only a subordinate role. For that reason the total local energy production $\sum_i I'_i$ is qualitatively the same as I'_2. The former is displayed in figure 4.46. Therefore, the symmetric basic flow becomes unstable for $\Gamma^e = 0.25$ due to the high shear between the recirculating bubbles and the almost

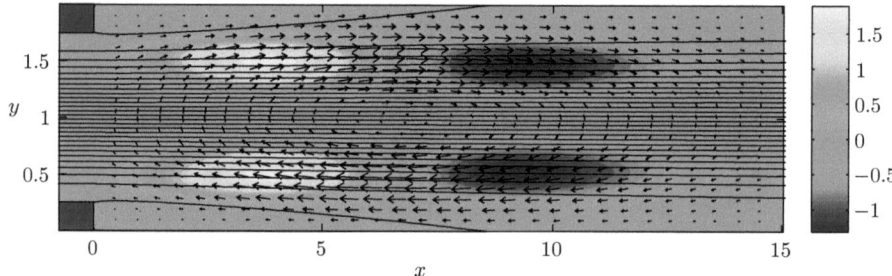

Figure 4.46.: Basic-state streamlines, critical velocity fields (arrows) and the total local energy production $\sum_i I'_i \approx I'_2$ for $\Gamma^e = 0.25$ at $\mathrm{Re}_c^{2D} = 1657.7$. Note that the scaling of the y axis is 2.5 times larger than that of x.

parallel jet in the bulk. The two-dimensional nature of the instability for this nearly parallel flow is consistent with Squire's theorem (Squire, 1933). Moreover, the velocity profile of the basic flow has a turning point, consistent with Rayleigh's theorem for inviscid flows.

Three-dimensional oscillatory instability for large expansion ratios $\Gamma^e \geq 0.713$

In the following, the linear stability of the asymmetric two-dimensional basic flow is considered.

Centrifugal instability for very large expansion ratios For very large expansion ratios, the basic state streamlines of the primary eddy are almost circular near its centre and they are only slightly strained. This can be seen in figure 4.47 for $\Gamma^e = 0.95$ at critical conditions, i.e. $\mathrm{Re}_c^0 = 423.48$ and $k_c^0 = 1.3031$.

In figure 4.47, the basic flow and the critical mode are shown in a cross-section $z = \mathrm{const.}$, in which the total local energy-transfer rate has its global maximum. The local flow is qualitatively

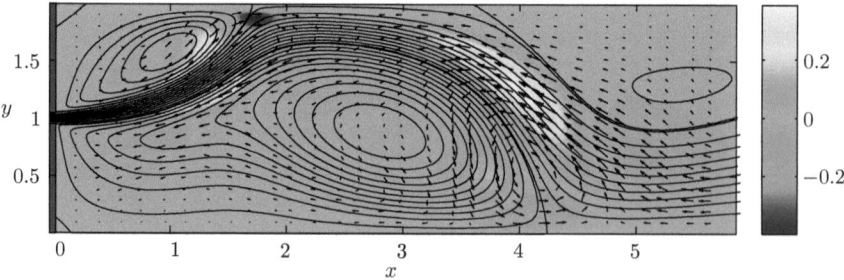

Figure 4.47.: Basic-state streamlines, critical velocity fields (arrows) and total local energy production $\sum_i I'_i$ for $\Gamma^e = 0.95$ and $\mathrm{Re}_c^0 = 423.48$ at $z = \mathrm{const.}$

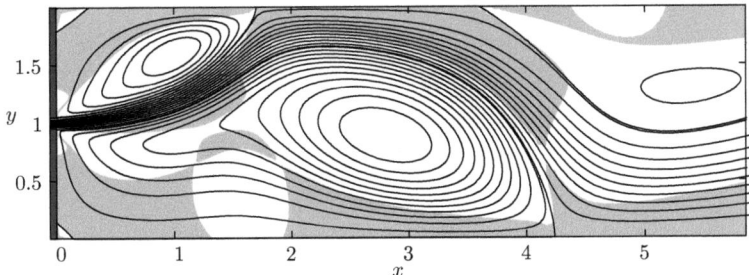

Figure 4.48.: Basic-state streamlines and grey regions where (4.3.6) is satisfied for $\Gamma^e = 0.95$ and $Re_c^0 = 423.48$.

the same near the two reattachment points $(x_t, y_t) \approx (1.69, 2)$ and $(x_b, y_b) \approx (4.20, 0)$ of the two separating streamlines, which originate near the sharp edges of the expansion. Also the local structure of the critical mode is similar near these points and consists of a vortex each, centred on the separating streamline slightly upstream of the reattachment point. There are also two regions of local energy transfer to the perturbation flow, which are stretched and located on the convex side of the jet, which is formed between the two separating streamlines of the basic flow. The local flow has curved streamlines and the velocity is rapidly decreasing outwards. Such flow conditions are prone to centrifugal effects and thus it is argued that the instability mechanism is of centrifugal type. This interpretation is supported by the criterion proposed by Sipp & Jacquin (2000), which is a reformulation of Bayly (1988)'s sufficient condition for centrifugal instability for two-dimensional, inviscid flows. Accordingly, the inviscid flow is centrifugally unstable if

$$\frac{|\boldsymbol{u}_0|\zeta}{R} < 0 \qquad (4.3.6)$$

is satisfied all along a closed streamline $\psi = \text{const}$. Here ζ stands for the vorticity of the basic flow and R for the local algebraic radius of curvature, given by

$$R = \frac{|\boldsymbol{u}_0|^3}{(\nabla\psi)\cdot(\boldsymbol{u}_0\cdot\nabla\boldsymbol{u}_0)}. \qquad (4.3.7)$$

Figure 4.48 shows in grey regions where (4.3.6) holds true. By comparing figure 4.47 with 4.48, it can be observed that the total local energy transfer is peaked and stretched parallel to the streamlines in just the grey regions where (4.3.6) is satisfied.

Moreover, one can hardly notice any difference between $\sum_i I'_i$ and I'_2 (not shown), as the term I'_2 is the most significant one in the integral sense (table 4.12). This observation is compatible with the interpretation that the flow becomes centrifugally unstable for very large expansion ratios $(1 - \Gamma^e) \ll 1$.

i	$\Gamma^e = 0.25$		$\Gamma^e = 0.5$		$\Gamma^e = 0.8$		$\Gamma^e = 0.95$	
	$\int_V I_i' dV$	$\int_V I_i'^{+} dV$	$\int_V I_i' dV$	$\int_V I_i'^{+} dV$	$\int_V I_i' dV$	$\int_V I_i'^{+} dV$	$\int_V I_i' dV$	$\int_V I_i'^{+} dV$
1	−0.0518	0.0201	0.0947	0.0997	0.0586	0.1082	−0.0194	0.0981
2	−0.1925	0.8744	0.2843	0.5507	0.6856	0.5452	0.6382	0.4827
3	0.0801	0.0095	0.0545	0.0683	0.0178	0.0957	0.0748	0.1459
4	1.1756	0.0960	0.5752	0.2813	0.2406	0.2626	0.3067	0.2747
$\int_{S_o}$	−0.0114		−0.0087		−0.0002		0.0000	

Table 4.12.: Global normalized energy production rates for selected expansion ratios for the asymmetric flow solutions Re_c^0.

It is interesting that the critical mode has a comparatively long wavelength of $\lambda_c^0 = 2\pi/k_c^0 = 4.8217$. Moreover, owing to the relatively low critical oscillation frequency $|\omega_c^0| = 0.0485$, the propagating wave travels only slowly in the spanwise direction with a phase velocity of $c^0 = |\omega_c^0|/k_c^0 = 0.0372$.

Note that the critical mode and the type of instability are the same as thoes detected in the backward-facing-step problem for $\Gamma^b > 0.9$ because the base flows are locally quite similar (see figures 4.10 and 4.11 for $\Gamma^b = 0.975$).

Elliptic instability for large expansion ratios As the expansion ratio is decreased, the critical mode changes continuously in a progressive manner and the centrifugal instability mechanism is phased out by an elliptic type of instability. The elliptic mechanism becomes dominant at around $\Gamma^e \lesssim 0.9$ and it also remains operative for smaller expansion ratios down to $\Gamma^e = 0.713$.

As a representative example, the flow structure is discussed for $\Gamma^e = 0.8$, where the asymmetric basic flow becomes unstable at $\mathrm{Re}_c^0 = 375.98$ with $k_c^0 = 1.3262$. In contrast to very large expansion ratios $(1 - \Gamma^e) \ll 1$, the primary vortex of the basic flow extends further downstream being quite elongated (see figure 4.49). The critical mode is confined within the primary eddy and is strongest in its centre. Figure 4.49 shows the spanwise plane $z = \mathrm{const.}$, in which the

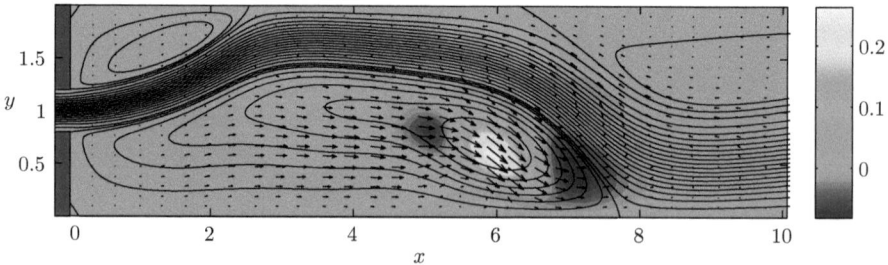

Figure 4.49.: Basic-state streamlines, critical velocity fields (arrows) and the total local energy production $\sum_i I_i'$ for $\Gamma^e = 0.8$ and $\mathrm{Re}_c^0 = 375.98$ at $z = \mathrm{const.}$

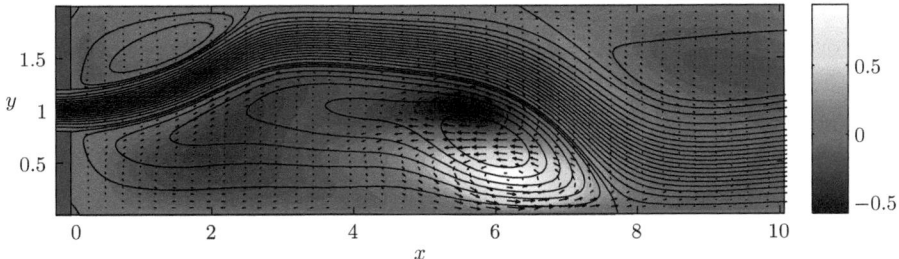

Figure 4.50.: Basic-state streamlines, vorticity of the critical mode (arrows) and $\tilde{w}$ (colour) for $\Gamma^e = 0.8$ and $\mathrm{Re}_c^0 = 375.98$ at $z = z_{\mathrm{amp}} + \lambda_c^0/4$.

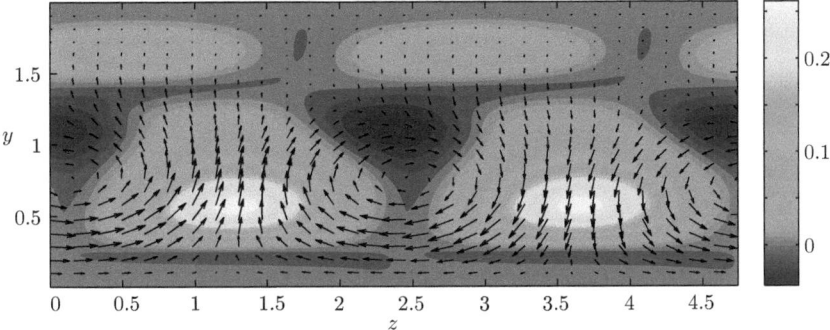

Figure 4.51.: Critical perturbation flow (arrows) and total local energy production $\sum_i I_i'$ for $\Gamma^e = 0.8$ and $\mathrm{Re}_c^0 = 375.98$ at $x = 6.0082$. The wave propagates to the right.

total local energy-transfer takes its maximum. In that plane, almost all the kinetic energy is produced in the centre of the elliptic vortex of the basic flow, which is a hallmark of an elliptic instability (Pierrehumbert, 1986; Sipp & Jacquin, 1998).

Figure 4.50 depicts the spanwise perturbation component $\tilde{w}$ in a plane z_{amp}, in which it has its maximum amplitude. This z_{amp} plane is shifted by a quarter of the critical wavelength $\Delta z = \pi/(2k_c^0)$. Figure 4.50 also shows the vorticity of the perturbation flow. The vorticity makes an angle of approximately $45°$ with respect to the major and minor axes of the elliptic streamlines, and is therefore aligned with the principal direction of strain. Such behaviour is another distinctive feature of the critical mode due to an elliptic instability (Waleffe, 1990).

At the margin of stability, the critical flow appears in the form of cellular structures, where the spanwise perturbation component $\tilde{w}$ vanishes periodically on equally spaced planes separated by $\Delta z = \lambda_c^0/2$. The cellular pattern is visible from one wavelength λ_c^0 in figure 4.51. The cut was taken at $x = 6.0082$ at the peak of the total local energy production. The absence of mirror symmetry planes at $z = \mathrm{const.} + \lambda_c^0/2$ indicates that the pattern is a travelling wave with the phase velocity $c^0 = |\omega_c^0|/k_c^0 = 0.0191$ and angular frequency $|\omega_c^0| = 0.0253$. The critical wavelength $\lambda_c^0 = 4.7377$ is approximately the same as for the centrifugal type of instability.

It is interesting to note that the instability mechanism in the range of Γ^e considered resembles that of the backward-facing-step flow for $0.709 \lesssim \Gamma^b \lesssim 0.9$.

Stationary instability for expansion ratios $\Gamma^e \leq 0.7$

Instability for $0.35 \leq \Gamma^e \leq 0.7$ In the parameter range $0.35 \leq \Gamma^e \leq 0.7$, all integral energy-production rates I_i' for $i = 1, 2, 3, 4$, are positive (see table 4.12 for $\Gamma^e = 0.5$) and thus contribute to the instability. The terms $I_2'^+$ and $I_4'^+$ are responsible for almost 83 % of all the positive energy production. However, the process I_1' exhibits the strongest maximum in the total local energy production, which makes the physical interpretation difficult.

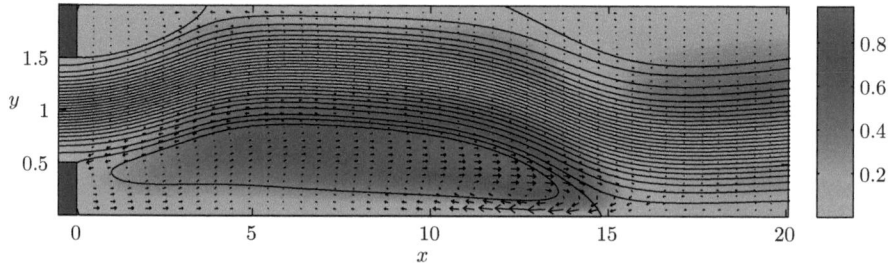

Figure 4.52.: Basic-state streamlines, spanwise vorticity of the critical mode (arrows) and $|\hat{\boldsymbol{u}}|$ (colour) for $\Gamma^e = 0.5$ and $\text{Re}_c^0 = 790.95$.

Figure 4.52 shows the streamlines of the base flow and the amplitude function of the critical mode $|\hat{\boldsymbol{u}}|$ (colour) for $\Gamma^e = 0.5$ at critical conditions, i.e. $\text{Re}_c^0 = 790.95$ with $k_c = 0.6339$. It also depicts the spanwise vorticity of the critical mode (arrows) in a plane, in which the spanwise perturbation component $\tilde{w}$ takes its maximum amplitude. It can be noticed that the amplitude of the perturbations as well as their vorticity are strongest on the downstream side of the primary vortex of the basic flow.

Figure 4.53 shows the critical flow pattern consisting of alternating streamwise streaks. It illustrates, moreover, that the total local energy-transfer, shown at its peak $y = 0.4147$, is strongly localized at around $x \approx 12.34$ and does not extend further upstream.

In figure 4.54, the total local energy-transfer rate $\sum_i I_i'$ and the above-mentioned amplification processes I_1', I_2' and I_4' are shown. All terms attain their extrema in the same planes $z = \text{const.}$ It can be seen from figure 4.54d that the critical mode gains most of its energy from I_4' near the separating streamline where the flow is strongly decelerated. In the same region, the flow is significantly stabilised through the process I_2' (figure 4.54c). Therefore, the term I_1' (figure 4.54b) contributes most to the maximum in the total local energy production $\sum_i I_i'$ (figure 4.54a). The

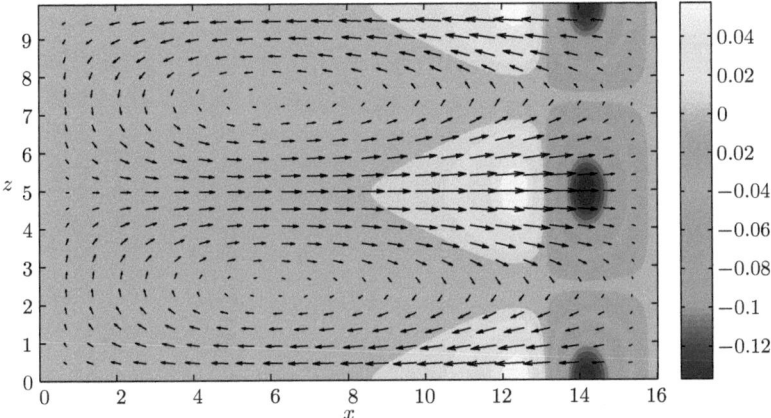

Figure 4.53.: Basic-state streamlines, perturbation flow (arrows) and the total local energy production $\sum_i I'_i$ at $y = 0.4147$ for $\Gamma^e = 0.5$ and $\text{Re}^0_c = 790.95$.

process I'_1 describes a cross-stream displacement of base flow momentum $\tilde{\boldsymbol{u}}_\perp \cdot \nabla \boldsymbol{u}_0$, feeding back on the perpendicular perturbation $\tilde{\boldsymbol{u}}_\perp$. In order for this process to have a destabilizing effect, the orientation of $\boldsymbol{u}_0$ must change perpendicular to it, which is satisfied for the converging streamlines of the basic state.

In light of the above, it is argued that in the parameter range $0.35 \leq \Gamma^e \leq 0.7$ the flow becomes unstable due to streamline convergence within the downstream side of the primary vortex (I'_1), in combination with flow deceleration near the reattachment point (I'_4) and an amplification process due to shear stresses near and between the primary and tertiary vortex (I'_2).

The critical mode and the mechanisms of instability for this instability are again very similar to the instability of the backward-facing-step flow for $0.4 \lesssim \Gamma^b < 0.709$. In fact, the basic flow topology near the reattachment point of the separating streamline is the same.

Shear instability for $\Gamma^e < 0.35$ For small expansion ratios, a relatively high Reynolds number is required to destabilise the flow. Under critical conditions the primary vortex of the basic flow is extremely stretched in the streamwise direction (figure 4.55). The steady and three-dimensional critical mode at $\text{Re}^0_c = 6434.2$ with $k^0_c = 0.1292$ for $\Gamma^e = 0.25$ is confined between the primary and the tertiary vortex of the basic state. The perturbation flow emerges in the form of a single vortex, which is centred at the channel centerline $y = 1$ and spanning the full outlet-channel width. The effect of a corresponding finite-amplitude perturbation would be an elongation and contraction of the primary and tertiary vortex, periodic in the z direction, which is accompanied by a straightening or bending of the jet between both separated vortices.

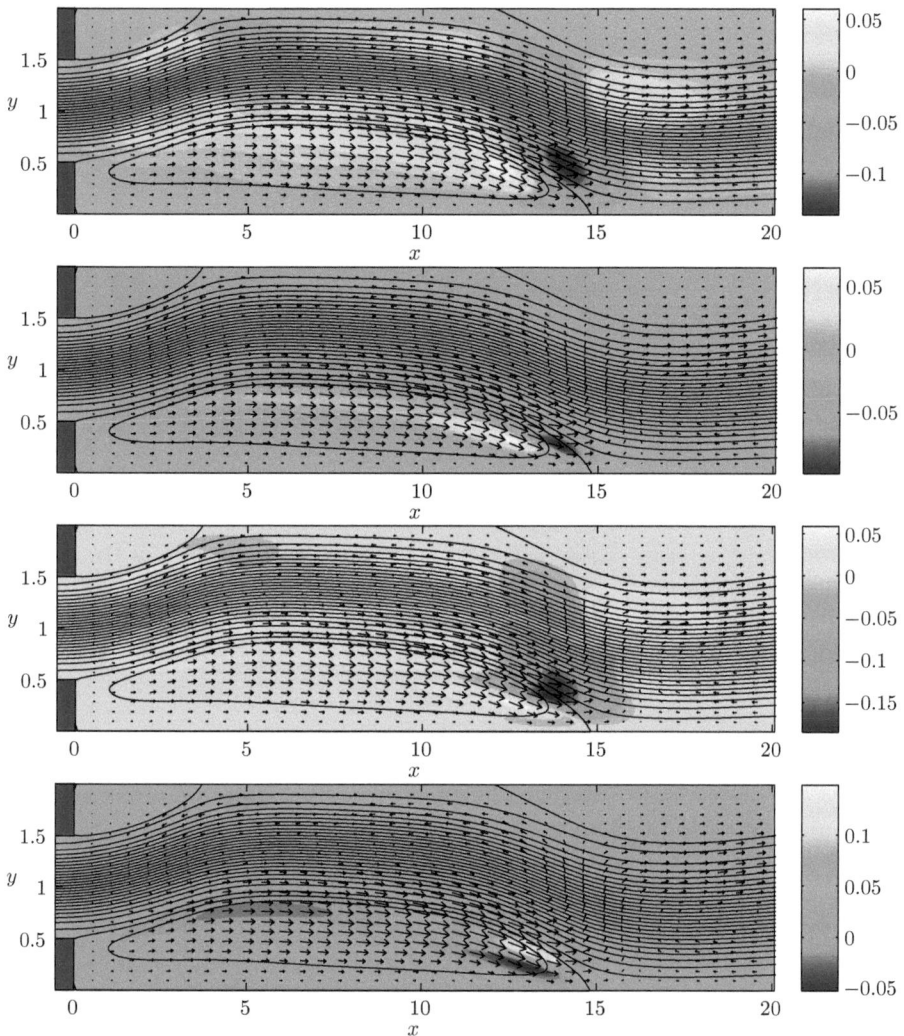

Figure 4.54.: Basic-state streamlines, critical velocity fields (arrows) and local energy production rates for $\Gamma^e = 0.5$ and $\mathrm{Re}_c^0 = 790.95$: (a) total local production $\sum_i I'_i$, (b) I'_1, (c) I'_2 and (d) I'_4.

From figure 4.55 it can be observed that the disturbances gain most of their energy in two regions, being located around $x \approx 23.31$. The peak value of the production in the region near the separation point of the tertiary vortex is marginally larger than the one near the reattachment point of the primary vortex. By inspection of $I_i'^+$ (table 4.12), one finds that $I_2'^+$ is responsible for about 87.44 % of all the positive energy production. Therefore, I_2' and $\sum I_i'$ are qualitatively the same. The same behaviour was found for the symmetry-breaking instability of the symmetric flow for $\Gamma^e = 0.25$ (section 4.3.3).

To demonstrate the streamwise distribution of the strength of the perturbation flow, figure 4.56 shows the amplitudes of the critical velocity field by components, integrated over y and averaged over z, i.e. $\mathrm{mean}_z \int |\tilde{\boldsymbol{u}}(x,y,z)| \mathrm{d}y$. It can be seen that the spanwise perturbation velocity component $\tilde{w}$ is rather weak. This result is consistent with the fact that the critical wavelength $\lambda_c^0 = 48.63$ is extremely long.

Owing to the similarities of the critical modes and its associated energy-transfer rates for the symmetric and the asymmetric basic flows (compare figures 4.55 and 4.46), the role of shear is mainly responsible for the instability for $\Gamma^e < 0.35$.

Comparison with experiments

The flow in a symmetric duct with an expansion ratio of $\Gamma^e = 0.5$ and a spanwise aspect ratio of $\Lambda = d/h_i = 8$ (depth-to-height of the inflow channel) was analysed by Cherdron et al. (1978) using flow visualizations and laser-Doppler anemometry measurements. Their experiments revealed that the flow was symmetric for Re = 150, but asymmetric for Re = 185. Moreover, it was reported that, for Re $\geq$ 800 onwards, the flow became very disturbed with large longitudinal velocity fluctuations. The experiments of Durst et al. (1993) for $\Gamma^e = 0.5$ and $\Lambda = 16$ predicted the symmetry-breaking bifurcation to occur at around $\mathrm{Re}_c^{2D} \approx 125$. Additionally, it was reported that the fully laminar regime reaches up to Re $\leq$ 610, which was the highest Reynolds number considered. Our global, temporal linear stability results predicted the

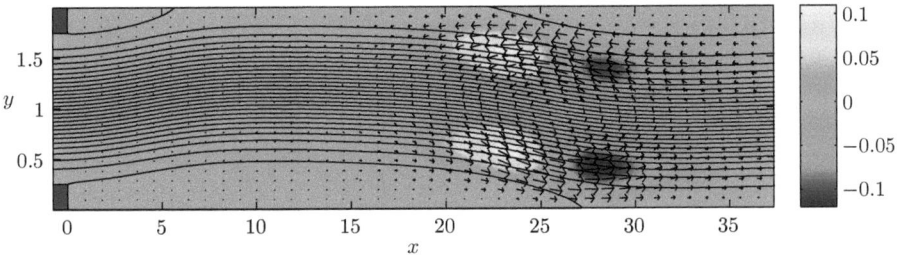

Figure 4.55.: Basic-state streamlines, perturbation flow (arrows) and the local energy production $I_2' \approx \sum I_i'$ for $\Gamma^e = 0.25$ and $\mathrm{Re}_c^0 = 6434.2$. Note the scaling of the x and y axes.

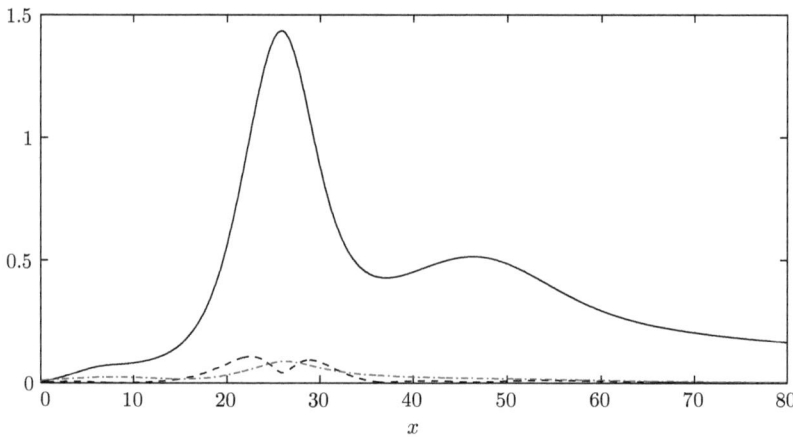

Figure 4.56.: Amplitudes of the perturbations, integrated over the y and averaged over the z direction mean$_z \int |\tilde{\boldsymbol{u}}(x,y,z)|\mathrm{d}y$ for $\Gamma^\mathrm{e} = 0.25$ and $\mathrm{Re}_c^0 = 6434.2$. The solid, the dashed and the dash-dotted (red) line represent the $\tilde{u}$, the $\tilde{v}$ and the $\tilde{w}$ component, respectively. The outlet is located at $x = 400$.

symmetry-breaking bifurcation to occur at $\mathrm{Re}_c^\mathrm{2D} = 216.76$. The discrepancy with the aforementioned experimental results of Cherdron et al. (1978) and Durst et al. (1993) might be due to the finite spanwise aspect ratios Λ. Apparently, Λ was too small to sufficiently prevent end effects in the experiments. This explanation is also supported by Tsui & Wang (2008) who showed that for a spanwise aspect ratio of $\Lambda \geq 24$ side-wall effects can be neglected and the flow can be regarded as two-dimensional at the centreline $z = 0$. Nevertheless, the stability boundary for the asymmetric flow solution $\mathrm{Re}_c^0 = 790.94$ is consistent with the experiments of Cherdron et al. (1978) with $\mathrm{Re}_c^\mathrm{Ch} \approx 800$.

The flow visualizations and laser-Doppler anemometry measurements of Durst et al. (1974) revealed that the flow is symmetric for Re = 84, but asymmetric for Re = 171 for $\Gamma^\mathrm{e} = 0.\dot{6}$ and $\Lambda = 27.5$. They also reported that the flow became very disturbed within the separation zones for Re ≥ 439.5. For approximately the same geometry, i.e. $\Gamma^\mathrm{e} = 0.\dot{6}$ and $\Lambda = 24$, experiments were carried out by Fearn et al. (1990). They could not detect any symmetric solution for Re ≥ 105. It was shown, moreover, that the flow remained steady up to Re ≤ 453 and the flow became unsteady as a consequence of three-dimensional effects for higher Reynolds number. These experimental results of Durst et al. (1974) and Fearn et al. (1990) agree quite well with our findings, yielding an exchange of stability at $\mathrm{Re}_c^\mathrm{2D} = 121.74$ and a loss of stability of the asymmetric flow for $\mathrm{Re}_c^0 \geq 454.83$.

4.3.4. Conclusion

The global, temporal linear stability of the two-dimensional, incompressible flow in a plane sudden expansion has been analysed for a quasi-continuous variation of the expansion ratio Γ^e. The influence of the in- and outlet-channel lengths on the stability boundaries has been analysed to obtain results, independent of these two external parameters. In addition, the flow stability for an asymmetric expansion with $\alpha = 0.15$ has been investigated. It turned out that the flow structure in the slightly non-symmetric geometry resembles the asymmetric flow solution in the symmetric geometry. Also the stability boundaries of these two basic flows are almost identical. Therefore, the main effect for a slightly asymmetric expansion is to shift the disconnected branches to higher Reynolds numbers.

The analysis has shown that the critical Reynolds numbers Re_c are monotonically decaying functions of the expansion ratio Γ^e. By using a definition of the Reynolds number based on a scaling, which depends only on the upstream conditions, Re_c^* scales linearly for very large step heights $(1 - \Gamma^e) \ll 1$. It is shown that the critical Reynolds number for the primary symmetric flow is significantly below that for the secondary asymmetric flow solution.

Our results confirm previous works showing that the primary instability is two-dimensional, stationary, and breaks the mirror symmetry with respect to the channel mid-plane. Using an *a posteriori* energy-transfer analysis it is demonstrated that the primary instability is due to a combination of flow deceleration and shear for large expansion ratios. For small expansion ratios, the role of flow deceleration diminishes and the instability is due to shear.

In addition to the primary instability, the secondary instability has been computed for the first time. The instability of the asymmetric flow solution is generally three-dimensional. Even though two-dimensional and time-dependent neutral modes exist (as anticipated by Cherdron et al. (1978) and Durst et al. (1993)), they are never the most dangerous ones. Typically, the neutral Reynolds number for two-dimensional time-dependent perturbations is at least twice the (three-dimensional) critical Reynolds number. For $\Gamma^e = 0.93$, for example, $Re_n^0(k=0) = 1167$ is obtained with $\omega_n^0(k=0) = 0.476$ and for $\Gamma^e = 0.95$, $Re_n^0(k=0) = 1023$ with $\omega_n^0(k=0) = 0.393$, respectively.

The three-dimensional instabilities are due to different physical mechanisms, depending on the expansion ratio. For small expansion ratios $\Gamma^e \leq 0.35$, the instability of the finite-amplitude two-dimensional asymmetric flow is stationary and similar to the primary instability. The critical three-dimensional, long-wave mode is almost mirror symmetric with respect to the mid-plane and it is caused by shear stresses. For moderate step heights $0.35 \leq \Gamma^e \leq 0.7$, the critical mode is still stationary, but the instability cannot be attributed to a particular type of energy-transfer process, because different mechanisms work together. In particular, the combined effects of flow deceleration near the reattachment point, an amplification process due to shear on both edges of the plane jet, and streamline convergence within the downstream region of the separated flow has a destabilizing effect. As the expansion is further increased, i.e. for $\Gamma^e \geq 0.713$, the critical mode changes from stationary to oscillatory. At $\Gamma^e \approx 0.8$ the disturbances gain their entire energy from the central region of the strained primary vortex. The energy-transfer characteristics and the structure of the critical perturbation flow indicate that the nature of the instability is of elliptic type. For even higher expansion ratios, the oscillatory critical mode and the energy-transfer characteristics change continuously until centrifugal forces prevail for very large step heights $(1 - \Gamma^e) \ll 1$. In this situation the local energy production and the perturbations are most pronounced in those regions where the plane jet of the basic flow is curved, just before its oblique impingement on the walls. There exist two zones with equivalent flow structures and energy-production zones on the concave side of the curved jet. The first zone is located in the curved jet just before its first impingement and the second zone is located in the curved separated jet just before the second impingement. These local flow conditions, and also the criterion of Sipp & Jacquin (2000), suggest that the flow becomes centrifugally unstable.

It is worth mentioning that the topology of the asymmetric critical modes ($\Gamma^e \gtrsim 0.4$) and their associated instability mechanisms are very similar to those studied in the backward-facing-step problem. These analogies seem to be due to the basic flows which are locally very similar in both systems.

The physical relevance of the global instability modes found is established by demonstrating the consistency with previous experimental findings. The numerical results are consistent with the experimental data regarding the primary symmetry-breaking bifurcation. Moreover, qualitative agreement is obtained for critical data of the secondary instability. In view of the remaining differences between numerical and experimental threshold data it would be worthwhile to carry out more accurate experiments in which the perturbing influence of the end walls is substantially minimized.

5. Summary and Outlook

The present study dealt with the primary instabilities in the backward-facing-step, forward-facing-step and plane sudden-expansion problems, respectively. The critical Reynolds and wave numbers were predicted numerically, at which the steady, incompressible and two-dimensional basic flow gets intrinsically three-dimensional. Great care was taken in obtaining stability boundaries independent of the lengths of the inlet and outlet channels. The geometric parameters of the above-mentioned systems were varied in a systematic way in order to cover a wide range of the parameter space, spanned by the Reynolds number Re and the geometric parameter Γ. Besides of that, the underlying instability mechanisms were clarified and characterized by means of an *a posteriori* kinetic energy-transfer analysis.

All the required numerical simulations were performed with a software package, which has been developed specifically for this doctoral thesis in MATLAB. The underlying equations were discretized using a finite-volume method being second-order accurate in space. The eigenvalues were computed with Arnoldi's method by applying shift-invert and Cayley transformations, respectively. All simulations run on a parallel computer for reasons of computing time and memory requirements.

A variety of new and interesting problems and open questions arise as a result of the current analysis. First of all, experimental results for validation are lacking for almost all the findings presented here. This study shows that a large spanwise aspect ratio is required for an accurate investigation of the primary instabilities, which should be taken into account in future experiments. But also fully three-dimensional numerical simulations including end walls in the spanwise direction would be of interest to study more systematically finite-size effects, caused by the presence of rigid walls. This could help to distinguish between sidewall effects and bulk-flow instabilities. Moreover, one could perform time-dependent simulations with the same boundary conditions as imposed here. It would be interesting to know which magnitude of the perturbations is required to destabilize the flow at the critical conditions. Thus the nonlinear dynamics of the bifurcation characteristics could be analysed and the total flow could be visualized. Future studies might also examine the transitional process from laminar to fully turbulent flow on the basis of Direct Numerical Simulations (DNS) and/or Large-Eddy Simulations (LES).

The present approach to the hydrodynamic stability analysis could also be applied to related geometries, such as the plane sudden-constriction or the periodically grooved channel problems.

The later system represents a combination of the backward-facing-step and forward-facing-step problems and the interactions of the vortices after and in front of the steps could be studied. Additionally, owing to the nature of this geometry, periodic boundary conditions could be used in the streamwise direction. Thus, if the step heights tend to zero, Tollmien–Schlichting waves should be recovered. In addition to the periodically grooved channel as studied by Ghaddar *et al.* (1986) and Amon & Patera (1989), one could combine the geometries of the plane sudden expansion and constriction as done by Takaoka *et al.* (2009).

In the current ansatz small perturbations were considered for investigating the primary instabilities. Alternatively, one could address higher Reynolds numbers and study the secondary flow. The flow, which arises after the primary instability of the basic flow is called the secondary flow, which might become unstable because of a secondary instability. Furthermore, a systematic perturbation analysis under the assumption of small disturbances could be carried out for the amplitudes of the perturbations to study the dominant nonlinear effects of higher orders.

Another open question is the existence of convective instabilities below the absolute stability boundaries as presented in the current work. Three-dimensional transient-growth analyses could be conducted in future studies in order to answer this question.

To conclude, one can state that the current study laid the foundation for the understanding of these fundamental model systems, i.e. the backward-facing-step, forward-facing-step and plane sudden-expansion problems. However, much more work has to be done in future studies to solve the fully nonlinear flow problem and to understand the above-mentioned open issues.

A. Derivation of the Reynolds–Orr Equation

Scalar multiplication of the linear perturbation equations (2.2.3) by $\tilde{\boldsymbol{u}}$ from the left-hand side followed by an integration over the volume $V = [0, 2\pi/k] \times \mathcal{D}$ yields (see e.g. Kuhlmann, 1999)

$$\frac{1}{2}\partial_t \left\langle \tilde{\boldsymbol{u}}^2 \right\rangle = \frac{1}{\mathrm{Re}} \left\langle \tilde{\boldsymbol{u}} \cdot \nabla^2 \tilde{\boldsymbol{u}} \right\rangle - \left\langle \tilde{\boldsymbol{u}} \cdot (\boldsymbol{u}_0 \cdot \nabla \tilde{\boldsymbol{u}}) \right\rangle - \left\langle \tilde{\boldsymbol{u}} \cdot (\tilde{\boldsymbol{u}} \cdot \nabla \boldsymbol{u}_0) \right\rangle - \left\langle \tilde{\boldsymbol{u}} \cdot \nabla \tilde{p} \right\rangle, \qquad (A.1)$$

where $\langle \ldots \rangle := \int_V \ldots \mathrm{d}V$. The first term $1/2\, \partial_t \left\langle \tilde{\boldsymbol{u}}^2 \right\rangle = \mathrm{d}E_\mathrm{kin}/\mathrm{d}t$ describes the rate of change of the total kinetic energy of the perturbations. In what follows the boundary conditions and the continuity equation are used to cast (A.1) into the Reynolds–Orr equation as given in (2.3.1).

The dissipation in the volume can be rewritten as

$$\begin{aligned}
\frac{1}{\mathrm{Re}} \int_V \tilde{\boldsymbol{u}} \cdot \nabla^2 \tilde{\boldsymbol{u}} \, \mathrm{d}V &= \frac{1}{\mathrm{Re}} \left[\int_S \underbrace{\tilde{\boldsymbol{u}} \cdot [(\boldsymbol{n} \cdot \nabla) \tilde{\boldsymbol{u}}]}_{=0} \, \mathrm{d}S - \int_V (\nabla \tilde{\boldsymbol{u}})^2 \, \mathrm{d}V \right] \\
&= -\frac{1}{\mathrm{Re}} \left[\int_V (\nabla \times \tilde{\boldsymbol{u}})^2 \, \mathrm{d}V + \int_{S_o} (\tilde{\boldsymbol{u}} \cdot \nabla)(\tilde{\boldsymbol{u}} \cdot \boldsymbol{n}_{S_o}) \, \mathrm{d}S \right] \\
&= \underbrace{-\frac{1}{\mathrm{Re}} \left[\int_V (\nabla \times \tilde{\boldsymbol{u}})^2 \, \mathrm{d}V + \int_{S_o} \tilde{u}\, \partial_x \tilde{u} + \underbrace{\tilde{v}\, \partial_y \tilde{u} + \tilde{w}\, \partial_z \tilde{u}}_{=0} \, \mathrm{d}S \right]}_{=:D},
\end{aligned} \qquad (A.2)$$

where S_o denotes the surface at the outlet and the normal vector $\boldsymbol{n}$ is defined in such a way that it points out of the computational domain $\mathcal{D}$. At the outlet boundary, it is defined by $\boldsymbol{n}_{S_o} = (1, 0, 0)^\mathrm{T}$. The other surface integrals at the inlet and at the solid walls vanish owing to the boundary conditions $\tilde{\boldsymbol{u}} = 0$. Note that the above formulation of the dissipation rate D is equivalent to the right-hand side of (2.3.2).

The surface integral (2.3.3) is obtained from $-\langle \tilde{\boldsymbol{u}} \cdot (\boldsymbol{u}_0 \cdot \nabla \tilde{\boldsymbol{u}}) \rangle$ as follows

$$\begin{aligned}
-\int_V \tilde{\boldsymbol{u}} \cdot (\boldsymbol{u}_0 \cdot \nabla \tilde{\boldsymbol{u}}) \, dV &= -\int_S \tilde{\boldsymbol{u}}^2 (\boldsymbol{u}_0 \cdot \boldsymbol{n}) \, dS + \int_V \tilde{\boldsymbol{u}} \cdot \nabla \cdot (\tilde{\boldsymbol{u}} \boldsymbol{u}_0) \, dV \\
&= -\int_{S_o} \tilde{\boldsymbol{u}}^2 u_0 \, dS + \int_V \tilde{\boldsymbol{u}} \cdot (\boldsymbol{u}_0 \cdot \nabla \tilde{\boldsymbol{u}}) \, dV + \int_V \tilde{\boldsymbol{u}}^2 \underbrace{(\nabla \cdot \boldsymbol{u}_0)}_{=0} \, dV \\
&= -\frac{1}{2} \int_{S_o} \tilde{\boldsymbol{u}}^2 u_0 \, dS.
\end{aligned} \quad (A.3)$$

Here the line of argumentation is the same as in equation (A.2).

For the pressure term $\langle \tilde{\boldsymbol{u}} \cdot \nabla \tilde{p} \rangle$ in (A.1), partial integration yields

$$\begin{aligned}
\int_V \tilde{\boldsymbol{u}} \cdot \nabla \tilde{p} \, dV &= \int_S \tilde{p} (\tilde{\boldsymbol{u}} \cdot \boldsymbol{n}) \, dS - \int_V \tilde{p} \underbrace{(\nabla \cdot \tilde{\boldsymbol{u}})}_{=0} \, dV \\
&= \int_{S_o} \tilde{p} \tilde{u} \, dS =: \tilde{p} \underbrace{\int_{S_o} \tilde{u} \, dS}_{=0} \quad \text{for} \quad \tilde{p} = \text{const.}
\end{aligned} \quad (A.4)$$

The above surface integral at the outlet vanishes because of the boundary conditions

$$\int_V \nabla \cdot \tilde{\boldsymbol{u}} \, dV = \int_S \tilde{\boldsymbol{u}} \cdot \boldsymbol{n} \, dS = \int_{S_o} \tilde{u} \, dS - \underbrace{\int_{S_{in}} \tilde{u} \, dS}_{=0} + \underbrace{\int_{S_w} \tilde{\boldsymbol{u}} \cdot \boldsymbol{n}_w \, dS}_{=0} = 0. \quad (A.5)$$

Since the surface integrals at the inlet and at the solid walls evaluate to zero due to the boundary conditions $\tilde{\boldsymbol{u}} = 0$, also the integral at the outlet must vanish so that the mass conservation is satisfied.

For the streamline coordinates, the perturbation flow $\tilde{\boldsymbol{u}}$ is decomposed into components parallel and perpendicular to the basic state $\boldsymbol{u}_0$. The decomposition (2.3.5) becomes numerically error-prone near stagnation points and shear layers, where $\boldsymbol{u}_0 \approx 0$. To avoid large errors associated with these regions, a small parameter $\check{\epsilon}$ is introduced

$$\tilde{\boldsymbol{u}}_\| = \frac{(\tilde{\boldsymbol{u}} \cdot \boldsymbol{u}_0) \boldsymbol{u}_0}{(\boldsymbol{u}_0 \cdot \boldsymbol{u}_0 + \check{\epsilon})}, \quad \tilde{\boldsymbol{u}}_\perp = \tilde{\boldsymbol{u}} - \tilde{\boldsymbol{u}}_\|. \quad (A.6)$$

If $\check{\epsilon}$ is selected as $\check{\epsilon} \leq 10^{-2}$, the unphysical energy production peaks are smoothed. This regularization did not effect the integral energy budget.

B. Jacobian-Free Newton–Krylov Approach

In the section 3.1 the Jacobian $J(x)$ of the stationary Navier–Stokes equations has been considered analytically. Building the Jacobian numerically is very time consuming and computationally unfeasible for large scale problems. By applying Krylov-subspace iteration, the Jacobian does not have to be formed explicitly, which is often referred to as the Jacobian-free Newton–Krylov (JFNK) method (Knoll & Keyes, 2004).

A wide variety of iterative schemes fall within the Krylov-subspace taxonomy. GMRES (generalized minimal residuals) and Bi-CGSTAB (stabilized bi-conjugate gradients) are two most frequently used types of Krylov-subspace methods, which are also applicable for solving nonsymmetric systems of equations and they have become quite popular recently in computational fluid dynamics. For details of GMRES and Bi-CGSTAB see Saad & Schultz (1986) and van der Vorst (1992), respectively. As these iterative schemes require the action of the Jacobian only in the form of matrix-vector products, the left-hand side of (3.1.1a) may be approximated by (Kelley, 2003)

$$J(x) \cdot \delta x \approx \frac{f(x + \epsilon\, \delta x/|\delta x|) - f(x)}{\epsilon} |\delta x|. \tag{B.1}$$

Note that scaling the direction in the forward-difference directional derivative (blue colour) is important to maintain superlinear convergence. Various options for choosing the perturbation parameter $\epsilon \ll 1$ are proposed in Knoll & Keyes (2004). Best results are obtained by setting $\epsilon = \sqrt{(1+|x|)\epsilon_{\text{mach}}}$, where ϵ_{mach} denotes the floating-point machine accuracy (typically $\approx 10^{-16}$ for 64-bit double precision). Equation (B.1) represents a first-order approximation to the Jacobian-vector product. Higher-order approximations are not frequently used in the JFNK approach, as at least one more function evaluation is required per matrix-vector product.

By applying an iterative scheme such as GMRES or Bi-CGSTAB for solving (B.1), a preconditioner is required because otherwise the method will not converge. Therefore, preconditioning plays a crucial role and is more important than the method itself. Finding a good preconditioner is not a trivial task as it depends on the problem and/or its generation may take longer than solving the system by itself. Generally speaking, a good preconditioner M should be better conditioned, more narrowly banded, and less expensive to build than $\boldsymbol{A}$ for the specific linear system $\boldsymbol{A} \cdot \boldsymbol{x} = \boldsymbol{b}$. A left preconditioner is applied in the following way

$$\boldsymbol{M}^{-1} \cdot \boldsymbol{A} \cdot \boldsymbol{x} = \boldsymbol{M}^{-1} \cdot \boldsymbol{b}, \tag{B.2}$$

whereas right-preconditioning is defined as

$$\boldsymbol{A} \cdot \boldsymbol{M}^{-1} \cdot \boldsymbol{y} = \boldsymbol{b} \quad \text{with} \quad \boldsymbol{y} = \boldsymbol{M} \cdot \boldsymbol{x}. \tag{B.3}$$

There is no general rule for choosing between a left and a right preconditioner, but one has to keep in mind that the iterations are terminated in a slightly different way as left-preconditioning changes the norm of the residual. Therefore, right-preconditioning is sometimes preferred by fitting better to the physical problem.

In the present work three different preconditioners have been analysed and tested: Stokes, SIMPLE and ILU (incomplete lower-upper) preconditioning. The first two of them are derived from the Navier–Stokes equations and are examples of the so called physics-based approach (Knoll & Keyes, 2004). The ILU preconditioner is independent of the underlying equations and represents a general preconditioner. The Stokes preconditioning as proposed by Edwards et al. (1994) considers only the diffusion term but did not work well, as the convection term plays a crucial role in the systems studied here. SIMPLE preconditioning as suggested by Vuik et al. (2000) was computationally feasible, but could not compete with the ILU preconditioner.

In the literature, there exits a great variety of ILU factorizations. The modified (MILU) and the Crout (ILUC) ILU decomposition turned out to be less effective than ILU(τ). Here the non-negative threshold τ specifies the drop tolerance, where all entries smaller than τ are set to zero. For $\tau = 0$ the complete LU factorization is computed (Saad, 2003). The time-critical parameter τ is mainly determined by the accuracy level, to which the underlying equations should be solved. Thus it should be of the same order of magnitude as the residual level (3.1.7), such as $\tau \approx \sqrt{\epsilon_{\text{mach}}}$. It is very important that the tolerance level of the iterative Krylov-subspace method (inner iteration) is at least two orders smaller than τ because otherwise the Jacobian-free Newton–Krylov scheme will not exhibit superlinear convergence. Note that reordering the system matrix $\boldsymbol{A}$, as described in the section 3.1, is an key element before applying any ILU factorization as it can speed up the calculations drastically.

To sum up, the terms $\boldsymbol{A}$ and $\boldsymbol{b}$ of the linear system $\boldsymbol{A} \cdot \boldsymbol{x} = \boldsymbol{b}$, which originate from Picard's linearization (3.1.5), are reordered according to the approximate minimum degree (AMD) algorithm. For the preconditioner, an ILU factorization is performed with a threshold level of $\tau \approx \sqrt{\epsilon_{\text{mach}}}$. Generating the preconditioner needed as much memory as solving directly the linear system with an efficient solver for sparse matrices, which converged, additionally, much faster than the iterative Krylov-subspace schemes. Therefore, the whole procedure as described in the appendix B was not used ultimately.

Bibliography

ALBENSOEDER, S., KUHLMANN, H. C. & RATH, H. J. 2001 Three-dimensional centrifugal-flow instabilities in the lid-driven cavity problem. *Phys. Fluids* **13**, 121–135.

ALLEBORN, N., NANDAKUMAR, K., RASZILLIER, H. & DURST, F. 1997 Further contributions on the two-dimensional flow in a sudden expansion. *J. Fluid Mech.* **330**, 169–188.

AMESTOY, P. R., DAVIS, T. A. & DUFF, I. S. 1996 An approximate minimum degree ordering algorithm. *SIAM J. on Matrix Analysis and Applications* **17**, 886–905.

AMON, C. H. & PATERA, A. T. 1989 Numerical calculation of stable three-dimensional tertiary states in grooved-channel flow. *Phys. Fluids* **1**, 2005.

ARMALY, B. F., DURST, F., PEREIRA, J. C. F. & SCHÖNUNG, B. 1983 Experimental and theoretical investigation of backward-facing step flow. *J. Fluid Mech.* **127**, 473–496.

BAI, Z., DEMMEL, J., DONGARRA, J., RUHE, A. & VAN DER VORST, H. 2000 *Templates for the Solution of Algebraic Eigenvalue Problems: A Practical Guide*. SIAM.

BARKLEY, D., GOMES, M. G. M. & HENDERSON, R. D. 2002 Three-dimensional instability in flow over a backward-facing step. *J. Fluid Mech.* **473**, 167–190.

BATCHELOR, G. K. 1967 *An Introduction to Fluid Dynamics*. Cambridge University Press.

BATTAGLIA, F., TAVENER, S. J., KULKARNI, A. K. & MERKLE, C. L. 1997 Bifurcation of low Reynolds number flows in symmetric channels. *AIAA* **35** (1).

BAYLY, B. J. 1986 Three-dimensional instability of elliptical flow. *Phys. Rev. Lett.* **57**, 2160–2163.

BAYLY, B. J. 1988 Three-dimensional centrifugal-type instabilities in inviscid two-dimensional flows. *Phys. Fluids* **31**, 56–64.

BEAUDOIN, J.-F., CADOT, O., AIDER, J.-L. & WESFREID, J. E. 2004 Three-dimensional stationary flow over a backward-facing step. *Eur. J.Mech. B/Fluids* **23**, 147–155.

BLACKBURN, H. M., BARKLEY, D. & SHERWIN, S. J. 2008 Convective instability and transient growth in flow over a backward-facing step. *J. Fluid Mech.* **603**, 271–304.

BLACKWELL, B. F. & PEPPER, D. W. 1992 Benchmark problems for heat transfer codes. In *Winter Annual Meeting of the American Society of Mechanical Engineers, HTD*, vol. 222. ASME.

BOTTARO, A., CORBETT, P. & LUCHINI, P. 2003 The effect of base flow variation on flow stability. *J. Fluid Mech.* **476**, 293–302.

BRENT, R. P. 1973 *Algorithms for Minimization without Derivatives*. Prentice-Hall.

CANTWELL, C. D., BARKLEY, D. & BLACKBURN, H. M. 2010 Transient growth analysis of flow through a sudden expansion in a circular pipe. *Phys. Fluids* **22**, 034101-1–034101-15.

CARMI, S. 1969 Energy stability of channel flows. *Z. Angew. Math. Phys.* **20** (4), 487–500.

CHANDRASEKHAR, S. 1961 *Hydrodynamic and Hydromagnetic Stability*. Oxford University Press.

CHERDRON, W., DURST, F. & WHITELAW, J. H. 1978 Asymmetric flows and instabilities in symmetric ducts with sudden expansions. *J. Fluid Mech.* **84** (1), 13–31.

CHIANG, T. P., SHEU, T. W. H., HWANG, R. R. & SAU, A. 2001 Spanwise bifurcation in plane-symmetric sudden-expansion flows. *Phys. Rev. E* **65**, 1–16.

CHIANG, T. P., SHEU, T. W. H. & WANG, S. K. 2000 Side wall effects on the structure of laminar flow over a plane-symmetric sudden expansion. *Computers & Fluids* **29**, 467–492.

CHIBA, K., ISHIDA, R. & NAKAMURA, K. 1995 Mechanism for entry flow instability through a forward-facing step channel. *J. Non-Newtonian Fluid Mech.* **57**, 271–282.

CHOMAZ, J.-M. 2005 Global instabilities in spatially developing flows: Non-normality and nonlinearity. *Annu. Rev. Fluid Mech.* **37**, 357–392.

CHUN, D. H. & SCHWARZ, W. H. 1967 Stability of the plane incompressible viscous wall jet subjected to small disturbances. *Phys. Fluids* **10**, 911–915.

CLIFFE, K. A., GARRATT, T. J. & SPENCE, A. 1993 Eigenvalues of the discretized Navier-Stokes equation with application to the detection of Hopf bifurcations. *Adv. Comput. Math.* **1**, 337–356.

CRIMINALE, W. O., JACKSON, T. L. & JOSLIN, R. D. 2003 *Theory and Computation in Hydrodynamic Stability*. Cambridge University Press.

CRUCHAGA, M. A. 1998 A study of the backward-facing step problem using a generalized streamline formulation. *Commun. Numer. Methods Eng.* **14**, 697–708.

DEISSLER, R. J. 1987 The convective nature of instability in plane Poiseuille flow. *Phys. Fluids* **30** (8), 2303–2305.

DENNIS, S. C. R. & SMITH, F. T. 1980 Steady flow through a channel with a symmetrical constriction in the form of a step. *Proc. R. Soc. Lond. A* **372**, 393–414.

DRAZIN, P. G. & REID, W. H. 1981 *Hydrodynamic Stability*. Cambridge University Press.

DRIKAKIS, D. 1997 Bifurcation phenomena in incompressible sudden expansion flows. *Phys. Fluids* **9** (1).

DURST, F., MELLING, A. & WHITELAW, J. H. 1974 Low Reynolds number flow over a plane symmetric sudden expansion. *J. Fluid Mech.* **64** (1), 111–128.

DURST, F., PEREIRA, J. C. F. & TROPEA, C. 1993 The plane symmetric sudden-expansion flow at low Reynolds numbers. *J. Fluid Mech.* **249**, 567–581.

EDWARDS, W. S., TUCKERMAN, L. S., FRIESNER, R. A. & SORENSEN, D. C. 1994 Krylov methods for the incompressible Navier–Stokes equations. *J. Comput. Phys.* **110**, 82–102.

ELOY, C. & LE DIZÈS, S. 2001 Stability of the Rankine vortex in a multipolar strain field. *Phys. Fluids* **13**, 660–676.

ERTURK, E. 2008 Numerical solutions of 2-D steady incompressible flow over a backward-facing step, Part I: High Reynolds number solutions. *Computers & Fluids* **37**, 633–655.

FEARN, R. M., MULLIN, T. & CLIFFE, K. A. 1990 Nonlinear flow phenomena in a symmetric sudden expansion. *J. Fluid Mech.* **211**, 595–608.

FERZIGER, J. H. & PERIĆ, M. 2002 *Computational Methods for Fluid Dynamics*. Springer.

FLETCHER, C. A. J. 1988 *Computational Techniques for Fluid Dynamics, Springer Series in Computational Physics*, vol. I. Springer.

FORTIN, A., JARDAK, M., GERVAIS, J. J. & PIERRE, R. 1997 Localization of Hopf bifurcations in fluid flow problems. *Int. J. Num. Meth. Fluids* **24**, 1185–1210.

GARTLING, D. K. 1990 A test problem for outflow boundary conditions - flow over a bachward-facing step. *Int. J. Num. Meth. Fluids* **11**, 953–967.

GHADDAR, N. K., KORCZAK, K. Z., MIKIC, B. B. & PATERA, A. T. 1986 Numerical investigation of incompressible flow in grooved channels. Part 1. Stability of self-sustained oscillations. *J. Fluid Mech.* **163**, 99–127.

GHIA, K. N., OSSWALD, G. A. & GHIA, U. 1989 Analysis of incompressible massively separated viscous flows using unsteady Navier–Stokes equations. *Int. J. Num. Meth. Fluids* **9**, 1025–1050.

GIANNETTI, F. & LUCHINI, P. 2007 Structural sensitivity of the first instability of the cylinder wake. *J. Fluid Mech.* **581**, 167–197.

GILBERT, J. R., MOLER, C. & SCHREIBER, R. 1992 Sparse Matrices in MATLAB: Design and Implementation. *SIAM J. Matrix Anal. Appl* **13**, 333–356.

GRESHO, P. M. 1991 Incompressible fluid dynamics: Some fundamental formulation issues. *Annu. Rev. Fluid Mech.* **23**, 413–453.

GRESHO, P. M., GARTLING, D. K., TORCZYNSKI, J. R., CLIFFE, K. A., WINTERS, K. H., GARRATT, T. J., SPENCE, A. & GOODRICH, J. W. 1993 Is the steady viscous incompressible two-dimensional flow over a backward-facing step at Re=800 stable? *Int. J. Num. Meth. Fluids* **17**, 501–541.

HASELGROVE, C. B. 1961 The solution of nonlinear equations and of differential equations with two-point boundary conditions. *The Computer Journal* **4**, 255–259.

HAWA, T. & RUSAK, Z. 2000 Viscous flow in a slightly asymmetric channel with a sudden expansion. *Phys. Fluids* **12** (9).

HAWA, T. & RUSAK, Z. 2001 The dynamics of a laminar flow in a symmetric channel with a sudden expansion. *J. Fluid Mech.* **436**, 283–320.

HAWA, T. & RUSAK, Z. 2002 Numerical-asymptotic expansion matching for computing a viscous flow around a sharp expansion corner. *Theor. Comput. Fluid Dyn.* **15**, 265–281.

HILL, D. C. 1995 Adjoint systems and their role in the receptivity problem for boundary layers. *J. Fluid Mech.* **292**, 183–204.

HOWELL, J. S. 2009 Computation of viscoelastic fluid flows using continuation methods. *J. Comput. Appl. Math.* **225**, 187–201.

HUERRE, P. & MONKEWITZ, P. A. 1985 Absolute and convective instabilities in shear layers. *J. Fluid Mech.* **159**, 151–168.

HUERRE, P. & ROSSI, M. 1998 Hydrodynamic instabilities in open flows. In *Hydrodynamics and Nonlinear Instabilities* (ed. C. Godréche & P. Manneville), chap. 2, pp. 81–294. Cambridge University Press.

JOHNSON, R. W. 1998 *The handbook of fluid dynamics*. CRC Press.

JOSEPH, D. D. 1976 *Stability of Fluid motions I*, Springer Tracts in Natural Philosophy, vol. 27. Springer.

KAIKTSIS, L., KARNIADAKIS, G. E. & ORSZAG, S. A. 1991 Onset of three-dimensionality, equilibria and early transition in flow over a backward-facing step. *J. Fluid Mech.* **231**, 501–528.

KAIKTSIS, L., KARNIADAKIS, G. E. & ORSZAG, S. A. 1996 Unsteadiness and convective instabilities in two-dimensional flow over a backward-facing step. *J. Fluid Mech.* **321**, 157–187.

KELLEY, C. T. 1995 *Iterative Methods for Linear and Nonlinear Equations*. SIAM.

KELLEY, C. T. 2003 *Solving nonlinear equations with Newton's method*. SIAM.

KERSWELL, R. R. 2002 Elliptical instability. *Annu. Rev. Fluid Mech.* **34**, 83–113.

KIM, J. & MOIN, P. 1985 Application of a fractional-step method to incompressible Navier–Stokes equations. *J. Comput. Phys.* **59**, 308–323.

KNOLL, D. A. & KEYES, D. E. 2004 Jacobian-free Newton–Krylov methods: a survey of approaches and applications. *Journal of Computational Physics* **193** (2), 357–397.

KUHLMANN, H. C. 1999 *Thermocapillary Convection in Models of Crystal Growth*, Springer Tracts in Modern Physics, vol. 152. Springer.

KUHLMANN, H. C., WANSCHURA, M. & RATH, H. J. 1997 Flow in two-sided lid-driven cavities: Non-uniqueness, instabilities, and cellular structures. *J. Fluid Mech.* **336**, 267–299.

LANDAHL, M. T. 1975 Wave breakdown and turbulence. *SIAM J. Appl. Math.* **28**, 735–756.

LANDAHL, M. T. 1980 A note on an algebraic instability of inviscid parallel shear flows. *J. Fluid Mech.* **98**, 243–251.

LANZERSTORFER, D. & KUHLMANN, H. C. 2012*a* Global stability of the two-dimensional flow over a backward-facing step. *J. Fluid Mech.* **693**, 1–27.

LANZERSTORFER, D. & KUHLMANN, H. C. 2012*b* Three-dimensional instability of the flow over a forward-facing step. *J. Fluid Mech.* **695**, 390–404.

LANZERSTORFER, D. & KUHLMANN, H. C. 2012*c* Global stability of multiple solutions in plane sudden-expansion flow. *J. Fluid Mech.* DOI: 10.1017/JFM.2012.184.

LEE, T. & MATEESCU, D. 1998 Experimental and numerical investigation of 2-D backward-facing step flow. *J. Fluids and Structures* **12**, 703–716.

LEHOUCQ, R. B. & SALINGER, A. G. 2001 Large-scale eigenvalue calculations for stability analysis of steady flows on massively parallel computers. *Int. J. Num. Meth. Fluids* **36**, 309–327.

LEHOUCQ, R. B. & SCOTT, J. A. 1997 Implicitly restarted Arnoldi methods and eigenvalues of the discretized Navier–Stokes equations. *SIAM J. Matrix Anal. Appl.* **23**, 551–562.

LEHOUCQ, R. B. & SORENSEN, D. C. 1996 Deflation techniques for an implicitly restarted Arnoldi iteration. *SIAM J. Matrix Anal. Appl.* **17**, 789–821.

MARINO, L. & LUCHINI, P. 2009 Adjoint analysis of the flow over a forward-facing step. *Theor. Comput. Fluid Dyn.* **23**, 37–54.

MATEESCU, D. & VENDITTI, D. A. 2001 Unsteady confined viscous flows with oscillating walls and multiple separation regions over a downstream-facing step. *J. Fluids and Structures* **15**, 1187–1205.

MEERBERGEN, K., SPENCE, A. & ROOSE, D. 1994 Shift-invert and Cayley transforms for detection of eigenvalues with largest real part of nonsymmetric matrices. *BIT Numerical Mathematics* **34**, 409–423.

MIZUSHIMA, J. & SHIOTANI, Y. 2000 Structural instability of the bifurcation diagram for two-dimensional flow in a channel with a sudden expansion. *J. Fluid Mech.* **420**, 131–145.

MOFFATT, H. K. 1964 Viscous and resistive eddies near a sharp corner. *J. Fluid Mech.* **18**, 1–18.

OL'SHANSKII, M. A. & STAROVEROV, V. M. 2000 On simulation of outflow boundary conditions in finite difference calculations for incompressible fluid. *Int. J. Num. Meth. Fluids* **33**, 499–534.

ORSZAG, S. A. 1971 Accurate solution of the Orr–Sommerfeld equation. *J. Fluid Mech.* **50** (4), 689–703.

PIERREHUMBERT, R. T. 1986 Universal short-wave instability of two-dimensional eddies in an inviscid fluid. *Phys. Rev. Lett.* **57**, 2157–2159.

PLOTKIN, A. & MEI, R. W. 1986 Navier–Stokes solutions for laminar incompressible flows in forward-facing step geometries. *AIAA J.* **24** (7), 1106–1111.

POLLARD, A., WAKARANI, N. & SHAW, J. 1996 Genesis and morphology of erosional shapes associated with turbulent flow over a forward-facing step. In *Coherent Flow Structures in Open Channels*, pp. 249–265. Wiley.

UR REHMAN, M., VUIK, C. & SEGAL, G. 2006 Solution of the incompressible Navier–Stokes equations with preconditioned Krylov subspace methods. *Tech. Rep.*. Delft University of Technology, Delft Institute of Applied Mathematics.

RUSAK, Z. & HAWA, T. 1999 A weakly nonlinear analysis of the dynamics of a viscous flow in a symmetric channel with a sudden expansion. *Phys. Fluids* **11** (12).

SAAD, Y. 2003 *Iterative methods for sparse linear systems*, 2nd edn. SIAM.

SAAD, Y. & SCHULTZ, M. H. 1986 GMRES: A Generalized Minimal Residual algorithm for solving nonsymmetric linear systems. *SIAM Journal on Scientific and Statistical Computing* **7** (3), 856–869.

SCHMID, P. J. & HENNINGSON, D. S. 2001 *Stability and transition in shear flows*. Springer.

SCHRECK, E. & SCHÄFER, M. 2000 Numerical study of bifurcation in three-dimensional sudden channel expansions. *Computers & Fluids* **29** (5), 583–593.

SEYDEL, R. 1994 *Practical Bifurcation and Stability Analysis: From Equilibrium to Choas*, 2nd edn. Springer.

SHAPIRA, M. & DEGANI, D. 1990 Stability and existence of multiple solutions for viscous flow in suddenly enlarged channels. *Computers & Fluids* **18** (3), 239–258.

SHTERN, V. & HUSSAIN, F. 2003 Effect of deceleration on jet instability. *J. Fluid Mech.* **480**, 283–309.

SIPP, D. & JACQUIN, L. 1998 Elliptic instability in two-dimensional flattened Taylor-Green vortices. *Phys. Fluids* **10**, 839–849.

SIPP, D. & JACQUIN, L. 2000 Three-dimensional centrifugal-type instabilities of two-dimensional flows in rotating systems. *Phys. Fluids* **12**, 1740–1748.

SOBEY, I. J. & DRAZIN, P. G. 1986 Bifurcations of two-dimensional channel flows. *J. Fluid Mech.* **171**, 263–287.

SOHN, J. L. 1988 Evaluation of FIDAP on some classical laminar and turbulent benchmarks. *Int. J. Num. Meth. Fluids* **8**, 1469–1490.

SPURK, J. H. 1997 *Fluid Mechanics*. Springer.

SQUIRE, H. B. 1933 On the stability of three-dimensional disturbances of viscous flow between parallel walls. *Proc. Roy. Soc. London* A **142**, 621–628.

STÜER, H. 1999 Investigation of separation on a forward facing step. PhD thesis, ETH Zürich.

STÜER, H., GYR, A. & KINZELBACH, W. 1999 Laminar separation on a forward facing step. *Eur. J. Mech. B/Fluids* **18**, 675–692.

TAKAOKA, M., SANO, T., YAMAMOTO, H. & MIZUSHIMA, J. 2009 Convective instability of flow in a symmetric channel with spatially periodic structures. *Phys. Fluids* **21**, 024105-1–024105-10.

THEOFILIS, V. 2000 Globally unstable basic flows in open cavities. In *6th AIAA/CEAS Aeroacoustics Conference*. AIAA 2000-1965.

THEOFILIS, V. 2003 Advances in global linear instability analysis of nonparallel and three-dimensional flows. *Prog. Aerospace Sci.* **39**, 249–315.

THEOFILIS, V. 2011 Global linear instability. *Annu. Rev. Fluid Mech.* **43**, 319–352.

THEOFILIS, V. & COLONIUS, T. 2011 Special issue on global flow instability and control. *Theor. Comput. Fluid Dyn.* **25**, 1–6.

THOMAS, L. H. 1953 The stability of plane Poiseuille flow. *Phys. Rev.* **91** (4), 780–783.

THOMPSON, J. F., WARSI, Z. U. A. & MASTIN, C. W. 1985 *Numerical Grid Generation: Foundations and Applications*. Elsevier North-Holland.

TREFETHEN, L. N., TREFETHEN, A. E., REDDY, S. C. & DRISCOLL, T. A. 1992 A new direction in hydrodynamic stability: Beyond eigenvalues. *Tech. Rep.* 92-71. ICASE, NASA contractor report 191411.

TREFETHEN, L. N., TREFETHEN, A. E., REDDY, S. C. & DRISCOLL, T. A. 1993 Hydrodynamic stability without eigenvalues. *Science* **261**, 578–584.

TSUI, Y.-Y. & WANG, H.-W. 2008 Side-wall effects on the bifurcation of the flow through a sudden expansion. *Int. J. Num. Meth. Fluids* **56**, 167–184.

VERSTEEG, H. K. & MALALASEKERA, W. 2007 *An Introduction to Computational Fluid Dynamics: The Finite Volume Method*, 2nd edn. Pearson.

VINOKUR, M. 1983 On one-dimensional stretching functions for finite-difference calculations. *J. Comput. Phys.* **50**, 215–234.

VAN DER VORST, H. A. 1992 Bi-CGSTAB: A fast and smoothly converging variant of Bi-CG for the solution of nonsymmetric linear systems. *SIAM Journal on Scientific and Statistical Computing* **13** (2), 631–644.

VUIK, C., SAGHIR, A. & BOERSTOEL, G. 2000 The Krylov accelerated SIMPLE(R) method for flow problems in industrial furnaces. *Int. J. for Num. Meth. Fluids* **33**, 1027–1040.

WALEFFE, F. 1990 On the three-dimensional instability of strained vortices. *Phys. Fluids* **2**, 76–80.

WESSELING, P. 2001 *Principles of Computational Fluid Dynamics*. Springer Series in Computational Mathematics.

WILHELM, D. 2000 Numerical investigation of three-dimensional separation in a forward-facing step flow using a spectral element method. PhD thesis, ETH Zürich.

WILHELM, D., HÄRTEL, C. & KLEISER, L. 2003 Computational analysis of the two-dimensional–three-dimensional transition in forward-facing step flow. *J. Fluid Mech.* **489**, 1–27.

WILLIAMS, P. T. & BAKER, A. J. 1997 Numerical simulations of laminar flow over a 3D backward-facing step. *Int. J. Num. Meth. Fluids* **24**, 1159–1183.

YANASE, S., KAWAHARA, G. & KIYAMA, H. 2001 Three-dimensional vortical structures of a backward-facing step flow at moderate Reynolds numbers. *J. Phys. Soc. Jap.* **70**, 3550–3555.

i want morebooks!

Buy your books fast and straightforward online - at one of world's fastest growing online book stores! Environmentally sound due to Print-on-Demand technologies.

Buy your books online at
www.get-morebooks.com

Kaufen Sie Ihre Bücher schnell und unkompliziert online – auf einer der am schnellsten wachsenden Buchhandelsplattformen weltweit! Dank Print-On-Demand umwelt- und ressourcenschonend produziert.

Bücher schneller online kaufen
www.morebooks.de

VDM Verlagsservicegesellschaft mbH
Heinrich-Böcking-Str. 6-8   Telefon: +49 681 3720 174   info@vdm-vsg.de
D - 66121 Saarbrücken   Telefax: +49 681 3720 1749 www.vdm-vsg.de

Printed by Books on Demand GmbH, Norderstedt / Germany